# The Life-centred Design Guide

Damien Lutz

ISBN: 978-0-6453266-5-9

Cover design by Damien Lutz.

*Lifecentred.design*

# Contents

# Figures

# Acknowledgement

The Life-centred Design Guide draws on the great work of sustainable, regenerative, and socially just practitioners past and present, such as Impossible, Vincit, The Design Council, The Ellen MacArthur Foundation, IDEO, Circulab, Disruptive Design, EcoDesign Circle, Kate Raworth and DEAL, Bruce Mau, Don Norman, Wholegrain Digital, Delft University of Technology, Biomimicry Institute, Karin Lidman and Sara Renström, Arturo Escobar, Dr. Renata M. Leitão, Dr. Lesley-Ann Noel, and many others, who have in various ways expanded the idea of what design should be. I'm very thankful for the inspiration which their work has given me, and I hope this book flows that inspiration forward to many others.

A heartfelt thank you goes to Susan Moylan-Coombs, Emma Ylivainio, Anhtai Anhtuan, Matt Cramsie, Dr. Dimeji Onafuwa, Dr S A Hamed Hosseini, and Ben Lowdon for their thoughtful advice that helped me shape the words for this guide.

I also acknowledge the Gadigal people of the Eora Nation as the Traditional Custodians of the Country where I live and work. I recognise their continuing connection to the land and waters and thank them for protecting this coastline and its ecosystems since time immemorial. I pay my respects to Elders of the past and present and extend that respect to all First Nations peoples.

# About the guide

Life-centred design, planet-centred design, planet-centric design, society-centred design, conscious design, 21st-century design, circular design, DesignX, respectful design, and humanity-driven design are all terms referring to an emerging design practice that eschews pure profit-driven goals for ones that give back to the planet and people they take from.

Life-centred design aims to achieve this by:

- Expanding human-centred product design to include consideration for the entire lifecycle, from material extraction, manufacturing, and shipping to use, end-of-use, and loops of reuse, repair, remake, and recycle
- Aligning product designers and businesses with global goals (Figure 1 - Life-centred Design Alignment)
- Protecting all peoples, all animals, and all planet along the lifecycle and regenerating them

For example:

- If product engineers use circular design to design their products with recyclable materials and to be easily repaired and disassembled, they can potentially shift tonnes of materials from landfill into loops of reuse that also reduce the need for more material extraction
- If digital designers employ sustainable digital design, behavioural design, and pluriversal design, they can reduce the carbon emissions by both business and users while also healing past marginalisation

While life-centred design strategies are mostly relevant to the design of physical products, this guide also recognises the hybrid physical/digital nature ('phygital') of today's creations and the ways to reduce the impacts of digital experiences on the physical world. Strategies for both are included in this guide to reflect this hybridisation and to educate designers of one discipline about the other.

Therefore, life-centred design as it is referred to in this guide, is:

*An adaptable, regenerative, and globally inclusive framework synching responsible businesses and designers with global goals to design products and services that minimise harm, re-nourish the planet, and foster fair, thriving, and diverse ways of being.*

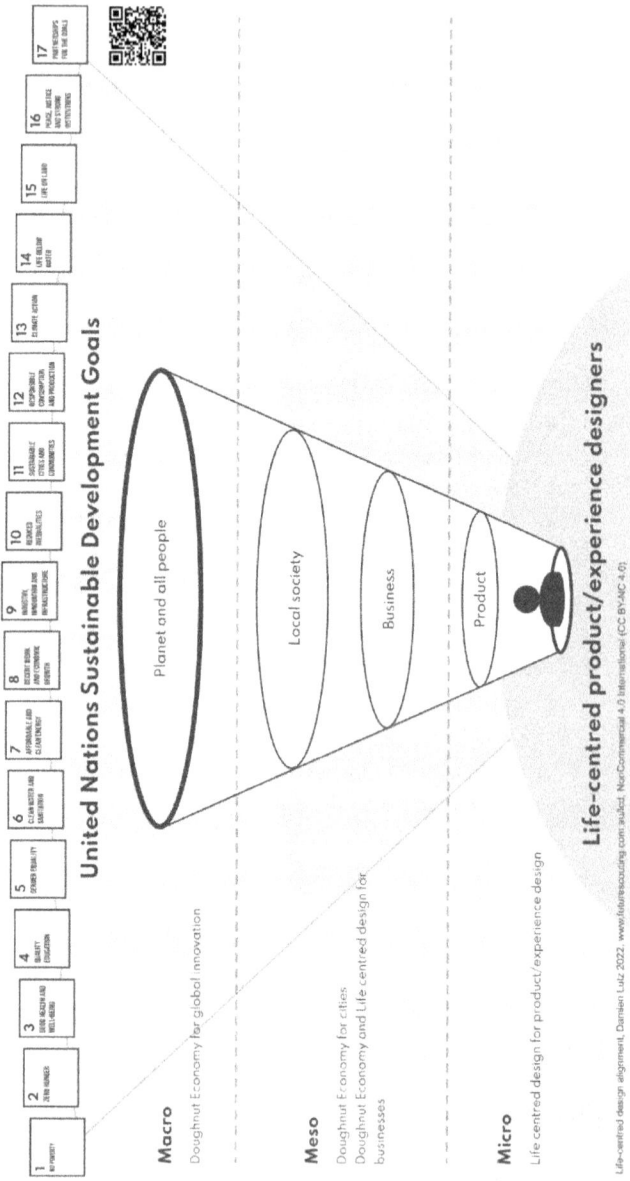

Figure 1 - Life-centred Design Alignment

## Who is this guide for?

This guide is for designers, service designers, and business decision makers with a passion for using their skills to address real issues, and for anyone interested in exploring transformational change from the ground up in the design, manufacturing, use, and reuse of physical and digital products.

## Why this guide?

As life-centred design is still emerging, awareness of it is low, those who practise it are few, the way it is practised varies, and digital and phygital strategies are nascent.

The *Life-centred Design Guide* is an introduction to current variations of life-centred design thinking and practice via a 'Future Snapshot', that combines 20+ life-centred design approaches, for the purposes of:

- Making discovery of life-centred design easier
- Creating a steppingstone to further educate others about life-centred design and the supporting practices
- Contributing to the emergence of life-centred design by generating discussion and experimentation
- Accelerating the transition to more regenerative and fairer local and global relationships between all people, animals, and planet

## What this guide is and isn't

This is not an engineering guide but a design thinking guide for the design of both physical and digital products and services.

The strategies, lenses, and tools provided in this guide are mostly experimental but based on the proven work of over 70 professional approaches and have been tested in either real or experimental environments.

Finally, the 'Future Snapshot' framework is not proposed as a 'final' framework—this should be created by a collaboration (and this is being addressed from an education perspective by *The Future of Design Education* initiative). Rather, this guide and framework are tools to compile the varying approaches, their common mindsets, their various sub-practices, and some informed speculation about life-centred design's future state, all in one introductory form, to ease the discovery of life-centred design and to inspire involvement from a wide variety of people and experience.

## How the guide is structured

- **Part 1**       **The story so far**—Background to the drivers of life-centred design's emergence and 20+ examples of current approaches
- **Part 2**       **The Future Snapshot**—The Future Snapshot framework explained
- **Part 3**       **The Practice**—A comprehensive set of strategies and methods to experiment with the Future Snapshot framework

## Bonuses

- Access to free tools and links to 100+ resources in *The Life Centred Design Resource Hub* (see next chapter)
- A short introduction to each of the 11 key design practices that combine in life-centred design with learning exercises (see the *Key design practices* chapter)

## How to use this guide

- **For a complete introduction** to life-centred design and how the Future Snapshot was created, read the guide end to end and complete the exercises using the tools from the resource hub (see next chapter). The guide and hub can then be used as an ongoing reference for further learning.
- **For a fast introduction**, read **The Story So Far** and then jump to **The Practice** and experiment with the strategies, lenses, and methods.
- **Just want to start experimenting?** Go straight to **The Practice.**

If you jump ahead, just make sure you read the rest of the book at some stage, as one of the key wisdoms in life-centred thinking is to know and respect the histories.

# The Life-centred Design Resource Hub

To accompany this guide, I created the online hub, lifecentred.design.

Sign-in to access:

- Tools for the **exercises** in this guide
- More tools for experimentation with **strategies and lenses**
- **Links** to more tools, courses, approaches, etc. for continued learning
- Reference material for the Future Snapshot framework

I have produced these tools and graphics for free use and as open source to encourage experimentation and contribution to the evolution of life-centred design. I just ask that you follow the licence guidelines in the hub.

Sign up/in to access.

For any feedback or suggestions on the material in this guide or in the hub, feel free to email me info@lifecentred.design

# Part 1
# -
# The story so far

# 1.1
# Designing
# today

Welcome, curious human.

You are the culmination of basic elements cycling for 2.8 million years through the 4.5-billion-year-old mix of oxygen, carbon, hydrogen, nitrogen, and sulphur that make up our planet—all originating from remnants of stars and massive explosions of galaxies from even further back in time.

In all that time and volatility, the simplest elements have remixed, recycled, and remade through circularity.

The regenerative way for us to exist in harmony with our world is in the universe's and our DNA.

But a large portion of humanity—mainly from the 'rich and developed countries'— have been consuming the resources of 1.7 Earths per year[1] through a destructive, polluting globalised industrialisation that uses massive amounts of non-renewable energy to ship continuous streams of refrigerators, antibiotics, throw-away furniture, and disposable wardrobes to prospering nations where they further enhance the lifestyles of the richest 20% and widen the wealth and equality gap, while producing so much $CO_2$ that the changing chemical composition of the atmosphere forced the planet's climate to shift into a new era—the Anthropocene era, the very first era that was not initiated by the planet, but by humans.

This human-driven destabilisation of the Earth is melting million-year-old glaciers and raising sea levels, forcing humans and animals to migrate en masse, water and food shortages to increase, and weather volatility to escalate to dangerous new norms.

The latest IPCC report confirms 'The world faces unavoidable multiple climate hazards over the next two decades with increase in the global warming of 1.5°C (2.7°F), some of which would be irreversible[2]'.

These increasingly unpredictable times are rendering the many traditional planning and foresight approaches driving decisions at every level as less and less reliable[3].

All this combines to create a frightening future for our children and any animals that are lucky to survive.

But we've heard all this many times before and in different ways—scientists and Indigenous Peoples have cried warnings for decades.

If this were a Netflix new release, or a timeless tale told around a fire, we'd be at the most exciting part—the Ordeal, the Abyss, where the Hero must face the truth about their own actions and survive a dangerous test or deep inner crisis to ensure the world continues to exist.

In many movies and stories, some transformational knowledge or technological Titanic-like marvel is introduced at the start. But by the time the Hero reaches the Ordeal, the technological marvel has been exposed as flawed, incompatible with nature, lacking enough life rafts for all, and dragging our Hero and everyone else into the abyss.

These stories tell us that knowledge or technology alone cannot equate humane intervention, that everyone's idea of utopia is different, and that our inventions and creations must respect their planetary source and be used in accordance with common human values that uplift all.

If we were all in a movie right now, *design* might be that flawed technological marvel, and the crisis our Hero must face would be to reinvent how they use design in more sustainable, regenerative, kind, and healing ways.

It could be argued that achieving the grand feats of the destructive global influences mentioned above suggests humans are also capable of the opposite—to foster a thriving harmony between all humanity, ecosystems, and other lifeforms.

Indigenous Peoples have lived with respect to the relationships between each other and the planet for thousands of years[4].

However, many immediately accessible online resources and academic journals about sustainable product and experience design don't refer to these origins of sustainable thinking. Instead, they refer to Euro-Western innovators of sustainable, circular, and systems thinking as 'pioneers' in their field. While true, the general exclusion of older wisdom and practices still alive in Indigenous Peoples today marginalises them and disconnects designers from their wisdom.

This disconnect from longer pasts, and its many perspectives, is just one persistent problem with modern design contributing to the world's 'wicked problems'.

Another problem is how this disconnect fuels the Global North's focus on constant growth as if it were the only way for humans to exist on planet Earth.

The Global North/Global South terms were first introduced in the 1970s as an alternative to terms such as 'Third World' to neutralise the idea that countries needed to 'qualify' for developed status to have as much value as developed countries.

Global North countries (Western Europe, Northern America, Australia, Israel, Japan, New Zealand, Singapore and South Korea) are wealthier, more democratic countries that export manufactured products. Global South countries (Central America, South America, Mexico, Africa, the Middle East, most of Asia, and others) are generally poorer with young institutions, and are often current or former subjects of colonialism.

The Global North/South concept is not a purely geographical divide as it can also refer to Global North powers within a Global South state and vice versa. This binary concept is challenged by some, but it does help articulate the divide between countries and peoples benefiting from globalised capitalism and those who are unfairly impacted by it.

While many modern innovations of industrialised nations have raised parts of the world out of recurring human famines into higher standards of living and education, increased human life expectancy, and improved poverty and sanitation for some, decades of the mass production of these innovations have forced others out of traditional living and into becoming a resource of abundant inexpensive labour and industry for the Global North's technological exports.

I, and many past and present Global North citizens, have contributed to this volatile moment hinging on how well all of humanity can handle the climate change and shifts of world power. We have been designing without a greater sense of connection and respect for natural systems and true human diversity—or, at least, without a sense of being able to do so.

But there are powerfully positive and healing impacts that all designers can start making now if they reinvent their practice to address three key problems of modern design:

- Designing with only humans in mind
- Designing for only a small part of the human spectrum
- Designing only for profit and constant growth

## 1. Designing with only humans in mind

With the invention of electricity in the 19th century, mechanical production evolved into mass production, leading to mass ownership. Greater consumer demand led to greater profits and prolific innovation, which in turn drove the mining machines to extract more minerals to make more products.

In this excitement, the design and manufacturing industry and its consumers established the acceptable limit to their responsibility of product design and manufacturing to the time from when a user purchases a product to when they discarded it—*take-make-waste*.

With automated and networked production in the mid-20th century, and the merging of electronics with consumer goods, more workers and consumers interacted with technology. In the 1980s, human-centred design arose to humanise this interaction, involving target-users in the design process to optimise the human experience. For several decades, human-centred design attuned experiences to customer's real needs, reduced frictions and cognitive load, and created more desirable products and experiences that made life better—but only better for businesses and their human target-users.

Separate courses for sustainable design also arose, however their decoupling from foundational product design courses effectively perpetuated the separation of designing for human commerce from designing for thriving people and planet.

The limited *take-make-waste* mentality ignored the sustained impacts on the environment and human communities along the product's entire lifecycle from extraction of its raw materials to its manufacturing, supply chains (often spanning multiple countries and cultures), to the use of the product, and finally until the product is discarded (which might include the breakdown of materials lasting hundreds of years).

When viewing products as part of a lifecycle system, we can no longer deny responsibility for releasing something into the living world as *a new ecosystem* of materials, people, and energy spanning the time and distance far beyond the target-user's use of the product and long after the producer stops producing and collecting profits.

Decades of designing without this regard for environmental and social impact, combined with the rapid expansion of consumers has far overextended humanity's limits of operation and fuelled the intensity of the climate crisis and widespread inequality.

## 2. Designing for only a small part of the human spectrum

While inclusivity and accessibility have been in design conversations for decades, and are always growing in consideration, they are yet to be comprehensively integrated as default considerations into product and experience design and education.

For example, with User Experience designers in high demand, new designers are often skilled up to address design problems, but accessibility and designing for diversity are often barely covered or offered as optional modules.

And, while some curriculums mention 'design in context', it is unclear how much is discussed about the historical negative impacts of design and the dangers of designers passing on biases that marginalise others, such as transferring discrimination into the design of powerful technology, like the AI of a chatbot, and, by proxy into all the human experiences with the chatbot.

It might be something as simple as designing the chatbot's personality to use overtly Euro-Western-centric names, references, and terminology to appeal to a target demographic, but which thereby subtly reinforces to anyone outside the Euro-Western culture that they are an 'other' and this world is not built for them.

In a more extreme case, in the US, an algorithm designed to calculate a score estimating the likelihood of a defendant re-offending was used by judges to decide verdicts and sentencing. The algorithm, however, drew from existing data that was embedded with the historical disproportionate targeting of minority communities, which thereby perpetuated that bias in its future calculations[5] and in the judges' verdicts and sentencing.

Designing for only the target user in mind ignores the impacts of a product's lifecycle upon 'invisible humans'—the people in the supply chain, or those communities affected by it. Forced labour, child labour, unfair pair, unsafe work conditions are all results of *take-make-waste* product design and manufacturing.

Also, mass production often sources materials from poorer countries for a much lower cost. This penetration of capitalism and purely human-centred/target-user thinking into these countries has disrupted those without stable institutions where power monopolies undermine, corrupt, and threaten poverty, inequality, and famine, driving the devastating impacts of conflict materials and further contributing to the Global North/South division.

### 3. Designing only for profit and growth

With mass production driving mass consumption since the 1920s, countries have been measuring their value and success according to the size of their economy, represented by their Gross Domestic Product value (GDP). In simple terms, the GDP is the value of goods and services produced by a country minus the value of the goods and services needed to produce them.

Clearly, this doesn't reflect the health of the country's society, environment, or relationships with other countries and the planet. It focuses on measuring things and quantities, rather than experience and quality, and with drastic implications. For example, a GDP is little affected by an airline improving its safety record, but a GDP can 'improve' if more planes crash because new fleets need to be redesigned and built[6].

The present and future disconnect created by the GDP mindset between a country's health, its people, and planetary resources is paralleled in product design's disconnect from lifecycle thinking. This focus on economic value has also influenced the individual purpose of many—to make as much personal profit as possible with little consideration for the planetary and human resources exploited.

### Planetary boundaries

From a scientific perspective, the coupling of constant development with unchained resource use threatens the planet's nine planetary boundaries, as proposed by *Stockholm Resilience Centre*[7].

These boundaries are the thresholds of our planet's key combined systems that are necessary for maintaining a habitable space for all life:

- Stratospheric ozone depletion
- Loss of biosphere integrity (biodiversity loss and extinctions)
- Chemical pollution and the release of novel entities
- Climate change
- Ocean acidification
- Freshwater consumption and the global hydrological cycle
- Land-system change
- Nitrogen and phosphorus flows to the biosphere and oceans
- Atmospheric aerosol loading

The original Planetary Boundaries Report suggested human activity had pushed at least four of these boundaries beyond their thresholds (climate change, biodiversity, land-system change, and biogeochemical flows). The report gave humanity until approximately 2028 before anything it did would stop having any significant influence on reducing the impacts of man-made climate change. This was only an estimate, and scientists still don't fully know what compounding effects may result from past, present, and future changes to these boundaries. Since the original report, however, the threshold for novel entities (chemical pollution) was crossed, and as recently as April/May 2022, water was added to the list, taking the total number of thresholds crossed to six.

The feedback loop happening between humanity and earth has made it clear that human-centred design's field of view needs to widen to include the true lifecycle of our products and their impact on the various human networks and natural ecosystems. The challenge for today is to transform the dominating linear system of *take-make-waste* into a circular system that sustains and regenerates both natural and human networks.

Yet, awareness of the connection between planetary, social, and economic health has never been higher.

More people than ever consider the environment and climate change to be the most important issues facing the world[8], driving renewed respect in the public for the importance of biodiversity and human wellbeing, and how they are connected. Consumers are expecting more transparency from brands and businesses to empower them to allow their values to drive more of their consumer decisions[9].

And the business world is responding.

The World Economic Forum is driving a full embrace of moving beyond profit-focused business to using 'capabilities and resources in cooperation with governments and civil society to address the key issues of this decade[10].' Some legacy actors, too, like mining companies, have set plans to become carbon negative. Billionaires are investing in scientists, activists, and other groups fostering transformational change. And not to forget the children warriors who have forgone school to challenge governments and the UN about mishandling their futures.

And so, we find ourselves at a critical inflection point.

Just as modern consumerism dreams of autonomous vehicles, flying cars, designer babies, and low-orbit space tourism near reality, so too do the splitting of the Doomsday glacier and unprecedented sea rise, both edging us closer to a soaring climate volatility and forced mass migration that will test all our systems, resilience, and compassion.

If this was all a Netflix new release, it could be the multi-million-dollar cliff-hanger scene of an alternate-reality sequel to *The Titanic*:

*Our Hero has been tossed from the sinking bow of the devastated passenger liner into the freezing North Atlantic waters. They splash in panic and desperately call for help. Screams echo off the hard water and groaning ship hull.*

*As the designer of the sinking vessel, our Hero is weighed down by guilt more heavily than the growing hypothermia.*

*'How could my grand design come to this?'*

*There seems no possible way for them to survive, their freezing, drowning doom inevitable...*

*When suddenly, a small boat appears, a female economist standing tall upon its bow, and she throws our Hero... a doughnut?*

# The Doughnut Economy

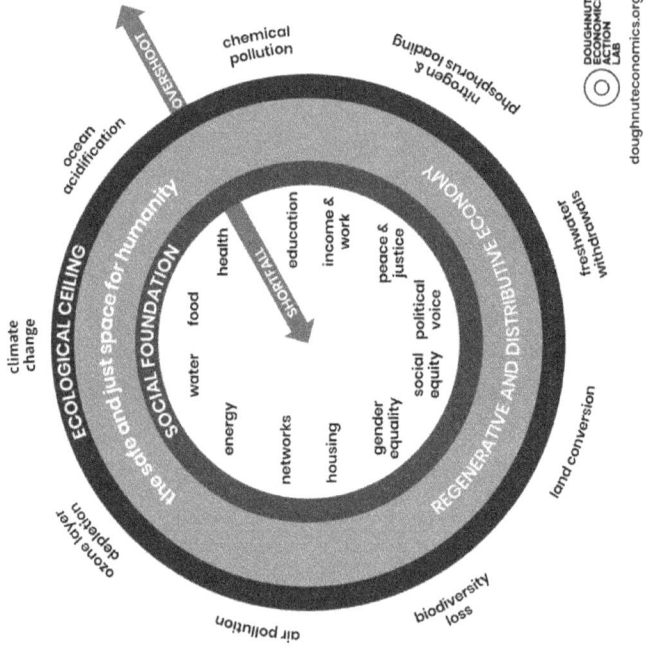

Figure 2 - The Doughnut Economy

## The Doughnut Economy

The Doughnut Economy was introduced by acclaimed economist Kate Raworth in her 2017 book *Doughnut Economics* as an alternative to the *take-make-waste* economy[11].

The Doughnut is currently used as a tool to assess development's impact on ecological and social issues, to plan for improvements, and to hold policymakers and citizens accountable (Figure 2 - The Doughnut Economy).

The Doughnut essentially represents a 'safe operating zone' between its inner and outer rings. The outer ring represents the nine planetary boundaries mentioned early that we should not overreach. The inner ring represents twelve 'basics of life' that we must ensure all peoples remain above to form a safe, just, meaningful, and thriving existence for all[12].

The space in between these two thresholds—the Doughnut—represents the target area where all activities should be focused in terms of outcomes.

Perhaps the Doughnut is a perfect symbol for these unpredictable times—a reflection of the gluttonous consumption that has contributed to the wicked problems of today, yet simultaneously reminiscent of the life ring thrown to drowning swimmers.

The Doughnut's twelve basics of life were derived from the social priorities specified in the United Nation's 2015 Sustainable Development Goals (SDGs)[13]:

- Sufficient food
- Clean water and decent sanitation
- Access to energy and clean cooking facilities
- Access to education and to healthcare
- Decent housing
- A minimum income and decent work
- Access to networks of information and to networks of social support
- Gender equality
- Social equity
- Political voice
- Peace and justice

Evolved over decades of work by the UN in conjunction with over 170 countries, the SDGs are 17 interconnected global goals that form 'a shared blueprint for peace and prosperity for people and the planet, at present and into the future'. The UN aims to mostly resolve these by 2030, which is around the same time the Stockholm Resilience Centre argues the world will run out of time to be able to reduce the impacts of man-made climate change.

The Doughnut Economics Action Lab (DEAL), co-founded by Kate Raworth, is part of the emerging global movement to rethink and reconfigure economic thinking by supporting regenerative and distributive projects of governments, communities, cities, businesses, and education. Its emphasis on sustainability and inclusiveness provides a promising framework to evolve modern design out of its three key problems.

Many global initiatives are adopting the Doughnut Model, and more and more businesses are incorporating the SDGs through the Environmental, Social, and Corporate Governance framework (ESG). ESG is a proposed metric released by the World Economic Forum in September 2021 for organisations to set and measure their sustainability improvements, support to societal movements, and governance. Put simply, organisations set commitments aligned with global goals, like the SDGs, from all three areas—environmental, social, and governance.

*This is important for life-centred design because it means the projects of product/experience designers working for a business with an ESG commitment are contributing to the shift toward a sustainable, regenerative, and fair world.*

But a 2021 study by DEAL reveal that no country is yet fully living within the boundaries of the Doughnut[14]. And the 2022 Circularity Gap Report revealed the world's total circularity was at only 8.6%, meaning a trillion tonnes of virgin materials are mostly going to waste and landfill. In 2018, the gap was 9.1%.[15]

Unfortunately, the challenges to adopting a more sustainable and regenerative economy are complex and many:

- Prioritising ESG commitments without compromising business goals
- Aligning stakeholders and partners on responsibility across the lifecycle
- True circular design requires extensive knowledge of all stages of a product lifecycle
- Current sustainable innovations and their progress can be minimal compared with what needs to happen
- Problems can be so big, expensive, and disruptive they become a political problem and taking so long that investors lose patience
- Some argue 'greening' capitalism is not enough, that we cannot heal the world while we are still polluting it, that a massive restructuring of our use of local and global resources is required[16]
- Those with much to lose from such a shift often hold much of the power to prevent needed changes

## Designing for the Doughnut

Supporting the sustainability aspect of the Doughnut Economy is circular design—designing products that keep their materials in loops of reuse to reduce the extraction of raw materials.

The best introduction to circular design is the butterfly diagram, adapted here in (Figure 3 - The Circular Economy). This was developed by the Ellen Macarthur

Foundation, a circular economy accelerator set up in 2010 by Dame Ellen Macarthur, a retired English sailor who broke the world record for solo circumnavigation of the globe in 2005.

Designing for a circular economy transforms *take-make-waste* into *reuse, repair, remake, and recycle* by designing goods to last longer and which can be easily disassembled so their materials can be remade or recycled into new goods, preventing the need for extraction of new materials.

Today, support for The Circular Economy is increasing worldwide with varying initiatives including waste reduction and business investment. Both new and long-established businesses are also making circular innovation:

- In 2017, Apple pledged to create all its products from recycled materials. By 2020, they had made the iPhone 11 product line circular by producing the device entirely from components of older Apple products17
- Inspired by the Global Fashion Agenda, Nike produced *Circularity: Guiding the Future of Design*, a guide for the sports fashion industry to create long lasting and circular products based on their own endeavours such as incorporating material offcuts into new products and refurbishing used shoes for resale
- Salubata, in Lagos, Nigeria, produce a modular shoe with a detachable and reusable sole and interchangeable uppers to reduce material waste and cost to consumer. Instead of buying multiple shoes, consumers can buy one sole and can vary the uppers with different colours and styles. As a bonus, the uppers are breathable and washable so wearing them doesn't require socks, reducing the energy and materials required to make socks!

## Life-centred design

In his 1971 book, *Design For The Real World*, Viktor Papanek opened with the provocative statement:

*"There are professions more harmful than industrial design, but only a few. And possibly only one profession is phonier. Advertising design...[18]".*

With his flair for precise and entertaining writing, Papanek challenged the limited consideration for design's negative impact on the world, and how the skills that created such harm where being taught to new generations of designers.

# The Circular Economy

Based on the Circular Economy model (Ellen MacArthur Foundation )

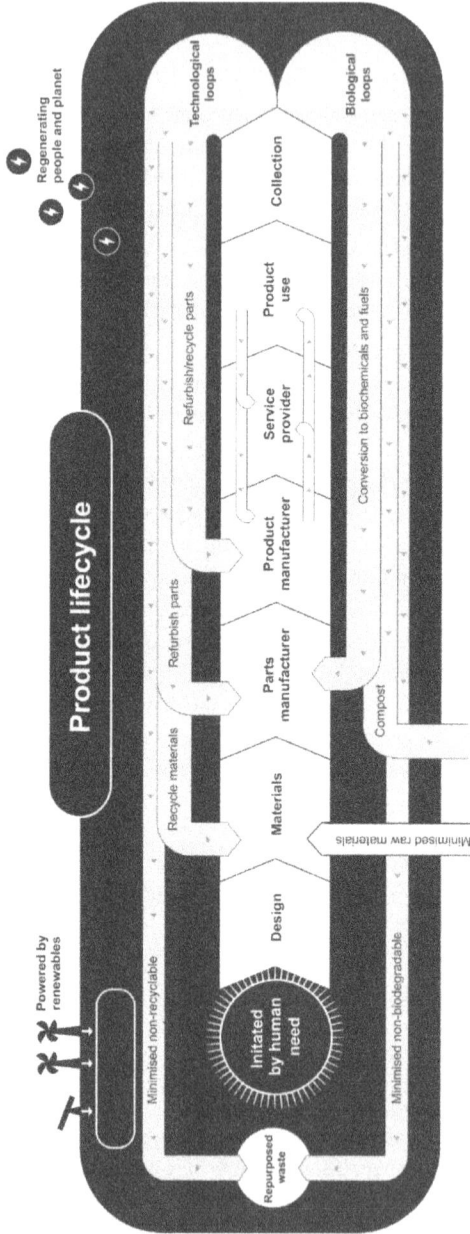

Figure 3 - The Circular Economy

*Design For The Real World* presented a blueprint for life-centred design that included product lifecycle thinking, biomimicry, designer responsibility, and an evolution of design education to reflect these expanded approaches. Papanek's perspectives were ridiculed by many at the time, but the book has since become one of the most widely read books about design. And today, various life-centred design approaches have emerged and are beginning to fulfil Papanek's vision. Some of these approaches utilise multiple design practices that have been practised for decades while others are new and just finding their legs, but all these approaches forgo a profit-only mindset in favour of a more collaborative, inclusive, holistic, and regenerative approach.

Yet, life-centred design is still finding place in the mainstream, meaning thousands of micro-level product designers and businesses are still enslaved to the fast-paced consumerism machine, encouraging people to buy stuff and do things that encourage the *take-make-waste* lifecycle. And the world's ever-increasing digital experiences are driving demand for electronic devices which is the fastest-growing sector for generating waste and which also drive demand for conflict materials.

Clarifying what this 21st Century thinking will foster the adoption of life-centred design at the micro-level and empower designers of physical and digital products to tackle wicked problems with global goals from the ground up.

As a bonus, learning life-centred design will expand product designers' abilities to implement design in a wider array of cultural, political, and socioeconomic situations. Don Norman argues for the importance of human-centred designers becoming more generalists[19] as they already possess the right skills to design impactful, sustainable, and just solutions that are:

- People-centred
- Identify and solve the right problem
- View everything as systems
- Make impactful iterations

Leyla Acaroglu, an Australian sustainability innovator and founder of the experimental knowledge lab, *Unschool*, also advocates for designers 'working within the multilevel perspectives' to not only make change in their designs, but also to be an active participant in making positive impacts on the world.

The blooming of life-centred design today acts like a reset of current product and digital design to include all our learnings to date—from human computer interaction design, human-centred design, design thinking, user experience design, and service design—and merges these with sustainable, regenerative, and inclusive thinking, while also protecting past wisdoms and future stakeholders.

In doing so, life-centred design reworks the three design problems into *How Might We?* statements to guide us toward our preferred futures:

- *How might we design with more than humans in mind, considering the impact of our creations upon the environment and animals across the entire product lifecycle, for now into the future?*
- *How might we design with all the spectrum of humanity, recognising alternate pasts, and protecting traditional wisdoms and different ways of living beyond the dominant Euro-Western-centric values?*
- *How might we design for thriving rather than for continuous growth?*

# 1.2
# Life-centred
# design approaches

To create the Future Snapshot, I researched over 20 approaches that expanded human-centred design to include the environment and/or social justice.

These included individuals, collectives, and organisations who theorised and/or practised a design process or educational approach that:

- Took a wider view than human-centred design (incorporating, for example, product lifecycle thinking, alignment with SDGS, design for social justice, etc.)
- Could be directly related to product design
- Had developed a unique and well-articulated approach

While this list of approaches is not an exhaustive one, it is hoped that it is broad and deep enough to establish a comprehensive summarised view of the scope of life-centred thinking today.

## Agricycle, Africa & Latin America

*Practice: Sustainability and a member-driven supply chain*
*Focus: Sustainability, Inclusivity, and Justice*
*agricycleglobal.com*

Activating in 6 countries and 16 languages across Africa and Latin America, Agricycle unites more than 35,000 farmers to produce up-cycled and ethically sourced products like fully traceable dried fruits (from tree to shelf) and fully sustainable grilling charcoals. By incorporating food that might never make it to market they reduce waste. The heart of their sustainable model is their decentralising, people-focused purpose—"We exist to eradicate extreme rural poverty." The model consists of three pillars:

- Appropriate and scalable technologies and systems
- Network-inclusive value chains that respects the rights of the poor, with each part generating revenue
- A focus on identifying and enabling markets

By selling their systems to people who don't own land to farm (often women, youth, and refugees) and including farmers in such a direct and distributed way and, Agricycle democratises access to the agricultural value chain which helps to eliminate extreme rural poverty.

## Bruce Mau, Principles for Massive Change, US

*Practice: Life-centred Design*
*Focus: Sustainability & Inclusivity*
*About the book: brucemaustudio.com/mc24*

*A summary of the principle's meanings:   medium.com/user-experience-design-1/bruce-maus-24-principles-for-life-centred-design-d026f35f5666*

Well-renowned Canadian designer, innovator, and educator, Bruce Mau leads the Massive Change Network, a consultancy exploring global innovation and sustainability. He began his career as a graphic designer and has since applied his design methodology to architecture, art, museums, film, eco-environmental design, education, and conceptual philosophy.

In his beautifully designed book, MC24, Mau shared his learnings from his 30+ years in the form of 24 principles for a more sustainable and just world.

**Principles:**

- Design is leadership: First inspire, Lead by design
- Begin with fact-based optimism
- Always search for the worst
- Seeing is believing
- We are not separate from nature
- Design for the Power Double Double
- Think forever: Design for perpetuity
- Design your own economy
- Sketch: Hey Everybody, let's fail!
- Think like you are lost in the forest
- Be whole-brain creative: it's a talent and a skill
- Compete with beauty
- Design for all the senses
- Rise above the noise
- Design for the time of your life
- Design the difference, not the object
- Design the platform for constant design
- Scale for impact
- Design the invisible
- Design the new normal
- Design what you do to tell your story
- New wicked problems demand new wicked teams
- Those who teach get out there and do
- Work on what you love

## Centre for Humane Technology

*Practice: Humane technology*
*Focus: Sustainability & Inclusivity*
*humanetech.com*

The Centre for Humane technology (CHT) champion a digital global infrastructure that supports well-being, democracy, and open-source software. Started by ex-Google Design Ethicist Tristan Harris, evolving from his talks addressing technology's misuse of distraction, CHT now train technology leaders, educate the public through media campaigns, and inform policy change to align with public good.

**Principles:**

- See in terms of human vulnerabilities
- Find and strengthen human brilliance
- We're constructing the social world
- Choosing is inevitable
- Bind growth with responsibility
- Design to enable wise choices
- Obsess over what really matters
- Nurture awareness

## Circulab, France

*Practice: Circular Design*
*Focus: Sustainability & Inclusivity*
*circulab.com*

Circulab began as Wiithaa in 2012, helping businesses up-cycle and design out waste long before circular design and circular economy were popular. They created the first Circular Canvas in 2014 and have since built a community of like-minded designers and experts, providing tools, training, and customised support.

Evolving into the strategy agency and design studio it is today, Circulab have also built the Circular Academy, an online learning platform for circular design. Their practices and teaching combine product lifecycle thinking, material mapping, systems thinking, and biomimicry. Through their project work and education, and in collaboration with their community, Circulab aims for three main goals[20]:

- Reduce resource consumption and optimise what is available
- Improve human and ecological resilience
- Regenerate living systems and biodiversity

**Principles and values:**

- Collective intelligence
- A holistic approach of innovation
- A smart use of resources
- Enthusiasm
- Accessibility
- Creativity
- Cooperation
- Consistency
- Selflessness

## The Circular Design Guide, UK

*Practice: Circular Design*
*Focus: Sustainability*

*circulardesignguide.com*

Produced in a collaboration between the Ellen MacArthur Foundation and global design company IDEO, this outstanding online resource is packed with methods, tools, examples, and advice for anyone wanting to start experimenting with circular design.

With its regenerative and product life-cycle approach, and inclusion of material mapping, systems thinking, and biomimicry, the Circular Design Guide provides tools and mindsets to expand human-centred product and business design into more sustainable and circular practices.

**Principles:**

- Design out waste and pollution
- Keep products and materials in use
- Regenerate natural systems

## Design Council, UK

*Systemic Design*

*Focus: Sustainability, Inclusivity, and Justice*

*designcouncil.org.uk/resources/guide/beyond-net-zero-systemic-design-approach*

The Design Council is a UK charity that champions the use of design to improve life and acts as the government's advisor on design. The Council was instrumental in promoting inclusive design and is well known for creating the Double Diamond framework for innovation.

In April 2021, the Design Council launched their Systemic Design Framework to help designers working on major complex challenges across different disciplines and sectors to place people and planet at the heart of design. Their six systemic design principles are accompanied by 4 proposed new design roles—System thinker, Leader and storyteller, Designer and maker, Connector and Convenor.

**Principles:**

- People and planet-centred
- Inclusive and welcoming difference
- Zooming in and out
- Collaborating and connecting
- Testing and growing
- Circular and regenerative

## Design Justice Network, US

*Practice: Design Justice*
*Focus:   Sustainability, Inclusivity, and Justice*
*designjustice.org*

The Design Justice Network connects designers through a network of nodes (meetup spaces) to share ideas about the application of the 10 *Design Justice Network Principles* that inspire designing for just, fair, and dignified experiences.

The 10 *Design Justice Network Principles*:

- We use design to sustain, heal, and empower our communities, as well as to seek liberation from exploitative and oppressive systems
- We centre the voices of those who are directly impacted by the outcomes of the design process.
- We prioritise design's impact on the community over the intentions of the designer
- We view change as emergent from an accountable, accessible, and collaborative process, rather than as a point at the end of a process
- We see the role of the designer as a facilitator rather than an expert
- We believe that everyone is an expert based on their own experience, and that we all have unique and brilliant contributions to bring to a design process
- We share design knowledge and tools with our communities
- We work towards sustainable, community-led and -controlled outcomes
- We work towards non-exploitative solutions that reconnect us to the earth and to each other
- Before seeking new design solutions, we look for what is already working at the community level. We honour and uplift traditional, Indigenous, and local knowledge and practices

## Disrupt Design, US

*Practice: Disruptive Design*
*Focus: Sustainability*
*disruptdesign.co*

Founded by Dr. Leyla Acaroglu, the New York based Disrupt Design uses its own systems-based sustainability design methodology to help businesses transition to sustainable, regenerative, and circular models. The Disrupt Design approach combines design thinking, sociology, environmental sciences, cognitive sciences, and systems thinking which they also teach via *The UnSchool of Disruptive Design*. Committed to open source, they also provide at least 20% of all their content and tools for free. Both the *Disrupt Design and Unschool* websites are bursting with a huge number of tools, inspiration, and learning.

**Principles summarised:**

- 3 knowledge pillars

  o Systems
  o Sustainability
  o Design

- 12 parts

  o Sustainability
  o Systems thinking
  o Making change
  o Research strategies
  o Gamification and game theory
  o Creative science and bias
  o Ethics and empathy
  o Language influence and effect
  o Systems interventions
  o Sustainable design and production
  o Project activation & amplification

## Don Norman, US

*Practice: 21st Century Design*
*Focus: Sustainability, Inclusivity, and Justice*
*interaction-design.org/master-classes/21st-century-design-with-don-norman*

Don Norman, a world leader in UX research and co-founder and Principal Emeritus of Nielsen Norman Group, is also known for introducing the term UX (as User Experience Architect of Apple) and fostering Human Centred Design.

To address how human-centred design can create or perpetuate problems for the environment, native and Indigenous cultures, and the many people who suffer by economic, racial, and class prejudices, Norman teaches *Design for the 21st Century*, a wider-view and more bottom-up approach where community members co-design with design professionals and experts.

The course explores the problems which can be resolved and avoided with better design, but that designers have no power—and that this can be addressed by designers thinking more broadly, thinking in systems, and learning more business thinking.

**Principles:**

- Always include the people being designing for
- Always include experts
- Avoid discussing a problem by its symptoms
- Muddle through via small, simple interventions
- Gain more influence

**Eco Design Circle Sustainability Guide, Baltic Sea Region**

*Practice: Ecodesign*

*Focus: Sustainability & Inclusivity*

*sustainabilityguide.eu*

The Ecodesign Learning Factory was developed from 2016 to 2021 as part of the EU-project *EcoDesign Circle* to address how businesses, designers, and design organisations in the Baltic Sea regions which could become more circular and sustainable.

*The Sustainability Guide*, one of several initiatives, is a knowledge platform to share the findings about ecodesign, circular economy, and sustainability, to allow design, business, and education to learn from each other. Focusing on the design phase, the platform is a great resource of inspiration, knowledge, methods, and examples, including product life-cycle approach, circular design, material choice, biomimicry, and systems design, to foster sustainable and circular business as well as social development.

**Principles:**

- Start with people
- Innovate
- Think in systems
- Convert to service

## Fairphone modular phones

*Practice: Sustainability, people, and planet*
*Focus: Sustainability, Inclusivity, and Justice*
*fairphone.com/en*

To tackle mobile phone e-waste, Amsterdam-based and B-Corp certified Fairphone encourages phone reuse and repair by selling modular smartphones and spare parts, and by offering repair tutorials.

They also advocate for a fairer electronics industry through practising responsible material sourcing and fighting for workers' welfare.

Fairphone use transparency about their journey through sharing reports and by being honest about not being perfect, that transitioning toward a more sustainable and fair business *is a journey* with challenges and obstacles.

## Fillipa K Fashion, Sweden

*Practice: Sustainability, Circularity*

*Focus: Sustainability, Inclusivity, and Justice*

*filippa-k.com/en/sustainability/circularity*

Swedish fashion house Fillipa K have translated the circular process into specific life-centred goals, broken down into achievable short-term and ideal long-term goals:

- Fibre use—Improve the class of all fibres, starting with 80%
- Traceability—Fully traceable and certified fibres, starting with 5 styles/products
- Longevity—Improve the use of garments by establishing and fostering a wearability KPI
- Circularity—Increase the resell, remake, and recycle of garments to reduce waste, starting with increasing the selling of second-hand garments in store
- Climate—Reduce annual emissions, starting with calculating emissions for select products
- Social—Ensure fair, safe, and non-exploited work conditions, starting with bias training for staff and working towards FairWear Foundation leader status

They also foster transparency by sharing their goals and their annual reports directly from their site.

## First Things First Manifesto, Global

*Practice: Conscious Design*
*Focus: Sustainability, Inclusivity, and Justice*
*firstthingsfirst2020.org*

First published in 1964, the First Things First manifesto was created by Ken Garland in collaboration with 20 other designers, photographers, and students, as a call for a more humanist and long-term design approach.

The manifesto outlines six principles calling for a re-focus of design's mindset to be more ethical and critical, regardless of scale or area of focus. The manifesto was rewritten and republished in 2000, 2014 and 2020. Evolving through different minds and hands over time, the manifesto is a living document allowing anyone to contribute to its next iteration.

**Principles summarised:**

- Challenge and examine the histories, processes, and ethics of design and develop new ways to design
- Understand that we are not outside of nature
- Commit to reconnecting design, manufacturing, distribution, and use of the things we design to the Earth — and all its inhabitants
- Support community-based efforts to advance and promote justice, healing, co-existence, and mutual respect
- Reverse the design profession's priorities in favour of more inclusive, empathetic, and engaged forms of action—a mind-shift that goes beyond sustainability—towards regeneration, exploration, and co-creation of a non-exploitative, non-appropriative set of social-environmental relations
- Direct design skills for the betterment of humanity towards a more ecological civilisation

## Future of Design Education, Global

*Practice: 21st Century Design*

*Focus: Sustainability, Inclusivity, and Justice*

*futureofdesigneducation.org*

The Future of Design Education initiative was founded in late 2019 by the Design Lab of the University of California and IBM's Global Design Group, with The World Design Organisation co-sponsoring. Its purpose is to evolve design education to address the modern needs of design—moving from a focus on object or singular experience design to more systemic, larger scale issues that can have complex socio-technical implications—and enabling designers to become advocates for social and environmental responsibility.

At time of writing, the initiative has established a structure of core, specialised, and elective study topics, and 20 themes and teaching contexts.

**Principles:**

- Act at the appropriate scale
- Anticipate the future
- Restore and sustain ecological balance
- Focus on people
- Reconcile competing priorities
- Strive for inclusivity
- Respect the importance of place and culture
- Accept accountability
- Follow a code of ethics

## Impossible, UK

*Planet Centric Design*

*Focus:   Sustainability & Inclusivity*

*impossible.com/pcd-methodology*

Impossible is a B Certified London-based innovation group and incubator that started as a gift economy platform before expanding into a technology focused design firm. Their Planet Centric Design aligns with the SDGs, and they use their own unique tools and methods to incubate and launch start-ups and improve existing services & businesses. Impossible align their clients' business with SDGs, identify the impacts to the wider lens of humans, planet, and animals, and then ideate solutions and mitigations.

Impossible applied their planet centric approach to their Bond Touch bracelets that utilise haptic tech to substitute physical touch and intimacy for long distance connections. After aligning their business purpose with the SDGs and refining them into goals, Impossible then set out on dual streams of work consisting of product feature updates and business and supply chain sustainability.

A case study of their method is shared in the "*Methods to get started*" chapter.

**Principles:**

- Systemic thinking, zooming in and out between micro and macro views
- Social behaviour change, inside the organisation and users
- Long-term thinking, backwards and forwards

## Karwai Ng and Will Anderson, UK & Europe

*Practice: Conscious Design*
*Focus: Sustainability & Inclusivity*
*medium.com/@willwai/conscious-design-manifesto-and-principles-a-will-wai-perspective-203b47cdb629*

'Will & Kar' are Berlin- and London-based design strategists who created principles in 2018 for what they termed 'conscious design'. They also produced a manifesto advocating for a wider design view of accountable and responsible awareness of the long-term consequences of design, and for continuous experimentation. Their approach includes using the iceberg framework for re-evaluating business value propositions and revealing unintended consequences of design.

**Principles:**

- Be conscious and prepared
- Look beyond the surface
- Feel uncomfortable
- Forget best-in-class
- Remember that design tools rust too
- Experiment with the process
- Challenge perspectives, do it in pairs

## NID Handbook, Netherlands

*Nature Inspired Design*
*Focus:   Sustainability & Inclusivity*
*natureinspireddesign.nl/handbook.html*

The Nature Inspired Handbook was created in 2014 out of a Circular Design Symposium run by the Delft University of Technology in Netherlands. Drawing in world pioneers of the circular economy, the Symposium aimed to inform designers and businesses on how to become more circular. Their approach is deeply inspired by biological systems using biomimicry.

While it's an extensive end-to-end process, with tools, instructions, and examples, it is a delightfully easy and interesting read, with the aim to realise product-systems that have a more positive impact on the environment. Their explanation of the workings of an Oak Tree as analogy for their principles is beautiful.

Principles:

- Adapt and evolve to changing conditions
- Waste equals food
- Use renewable energy
- Be locally attuned and responsive
- Be resource efficient
- Integrate development with growth

## Nike Circularity Guide, US

*Practice: Circular Design*
*Focus: Sustainability, Inclusivity, and Justice*
*nikecirculardesign.com*

Nike created their circularity guide for the footwear fashion industry in collaboration with the London University of the Arts, with inspiration and insights from Global Fashion Agenda, and the Ellen MacArthur Foundation, to help make a shift in their own industry.

The guide is a workbook to be used by designers and businesses through each design stage, with examples and prompts on how to be more circular. The examples provide excellent insight into practical business application of circular thinking which can otherwise be an exciting but overwhelming collection of principles, tools, and methods.

Sharing their journey towards a 'zero carbon and zero waste of future sport', Nike reports the way they recycle and refurbish 'gently used' shoes to resell at a lower price, extending the value of the materials used in the shoe and potentially reducing purchasing of new shoes.

They also incorporate recycled materials (carpet, used fish nets, and plastic bottle waste) to reduce material waste through better processes and reuse of offcuts, and through sourcing 100% of their cotton from Better Cotton, an initiative that advocates for sustainable farming and improved working conditions and standards of living for farmers[21].

**Principle focuses:**

- Material choices
- Cyclability
- Waste avoidance
- Disassembly
- Green Chemistry
- Refurbishment
- Versatility
- Durability
- Circular Packaging
- New business models

## Patagonia, US

*Practice: Social Responsibility*
*Focus: Sustainability, Inclusivity, and Justice*
*patagonia.com/social-responsibility*

A great example of a life-centred business is Patagonia, an American outdoor clothing company with hundreds of stores in over 10 countries across 5 continents, and factories in 16 countries. They've aligned their organisation with global goals by utilising their product value, lifecycle connections, and areas of influence to transform from a business just profiting from selling clothes to also championing the many following initiatives:

- Created their manifesto about becoming an antiracist company
- Support fair work conditions for apparel workers via a social-responsibility program that analyses and manages the impacts of the business it has on the workers and communities in the supply chain
- Utilise Regenerative Organic Certified™ Programs that support people and animals 'working together to restore the health of our planet' by improving soil health and reducing greenhouse gas emissions
- Pledged 1% of sales to the preservation and restoration of the natural environment
- Created Worn Wear to encourage the reuse of unwanted clothing and keep the materials in use
- Provide financial and networking support to environmental action groups
- Use transparency to share supply chain information, so customers know where and how their clothes are made
- Set climate-specific goals to reduce carbon emissions from across the entire supply chain
- Share their ethical footprint to remain transparent and accountable

## Sentient Future Lab, China/Japan-based

*Practice: Life-centred Design*

*Focus: Sustainability & Inclusivity*

*medium.com/@sentientcollective/10-principles-of-life-centered-design-948ac1da7a56*

Sentient Future Lab is a Shanghai and Tokyo based global design collective of designers, strategists, educators, filmmakers, data scientists, and engineers dedicated to solving world. Their founder, Johnathyn Owens, shared *10 Principles of Life-centred Design* for a more sustainable, inclusive, and open-source approach to design.

**Principles summarised:**

- Design with the full picture in mind
- Think about the future, just as much as about now
- Design is for all, not just for those who can 'afford it'
- The bottom line is necessity, not cost
- Design to last, not to fail
- Design down to the last detail
- Design symbiotic with nature
- Design in intelligence
- Design in humaneness
- Design is as few things as possible

## Vincit, Finland

*Practice: Planet Centric Design*
*Focus: Sustainability & Inclusivity*
*planetcentricdesign.com*

Vincit is a software development and digital transformation company using sustainable and systemic design, focusing on reducing the environmental impact of digital services. They developed their own Planet Centric Design methodology for 'designing products and services that do not harm the planet'.

Their complete end-to-end approach includes 19 methods, addressing attitudes toward climate change before shifting into impact-assessment, vision setting a concept development process 'infused with planet-centricity', business model flipping, and winning over stakeholders.

**Principles:**

- Responsible
- Systemic
- Transparent
- Wider lens
- Desirable Futures
- Sustainable Digitisation

## Wholegrain Digital, UK

*Practice: Sustainable Web Design*

*Focus: Sustainability & Inclusivity*

*wholegraindigital.com*

B-Corp certified Wholegrain Digital is a London-based sustainable web design business that also actively advocates for a sustainable and ethical internet.

Through the promotion of circularity and regeneration, creating low carbon websites, and using web hosting powered by renewable sources, they aim to shift the industry towards a zero-carbon future and create a better internet for people and planet.

They developed the first ever methodology for calculating the energy and carbon emissions of web pages, and they contributed to the Sustainable Web Design Manifesto which states six simple principles for reducing environmental damage and developing a healthy and fair global and local society

Cofounder and Managing Director Tom Greenwood also authored the Sustainable Web Design book, an excellent resource with practical tips to reduce the carbon emissions of web sites.

To further the societal improvement aspect of their life-centred work, Wholegrain Digital apply their sustainable web design expertise to assist other B-Corp certified organisations to promote their ethical messages.

Their principles promote an open, honest, clean, efficient, regenerative, and resilient industry.

# Part
# 2
-
# The Future
# Snapshot

# 2.1
# The framework

As you can see, each approach has its own unique perspectives, focus, methods, and values.

**Commonalities of approaches:**

- Considering entire product lifecycle
- Considering impact on environment
- Connecting product and service design with wider environmental and/or social goals

**Variations of approaches:**

- Designing for circularity—Some approaches incorporate circular design strategies
- Focusing on inclusivity as much as sustainability—Some approaches focus on sustainability without a clear equal focus on inclusivity or social justice
- Choice of supporting design practices—The varying approaches utilise different mixes of design practices, such as circular design, systems thinking, biomimicry, futures studies, and others
- Aligning projects specifically with SDGs—Some approaches use SDGs as goals while others use a generalised sustainability approach to determine project goals

**Issues addressed minimally or not at all:**

- Fostering social justice—Beyond target-user inclusivity, recognising and neutralising injustices along the supply chain while also aiming to heal
- Being aware and critical of design processes—Being mindful of the negative impacts of design and the biases and privileges of the designer
- Designing for animals—Design considerations and tools specific to animals' unique abilities, needs, and ways of existing
- Design strategy detail for digital and phygital design problems

# Life-centred design Future Snapshot

A speculation of life-centred design from a merging of current approaches

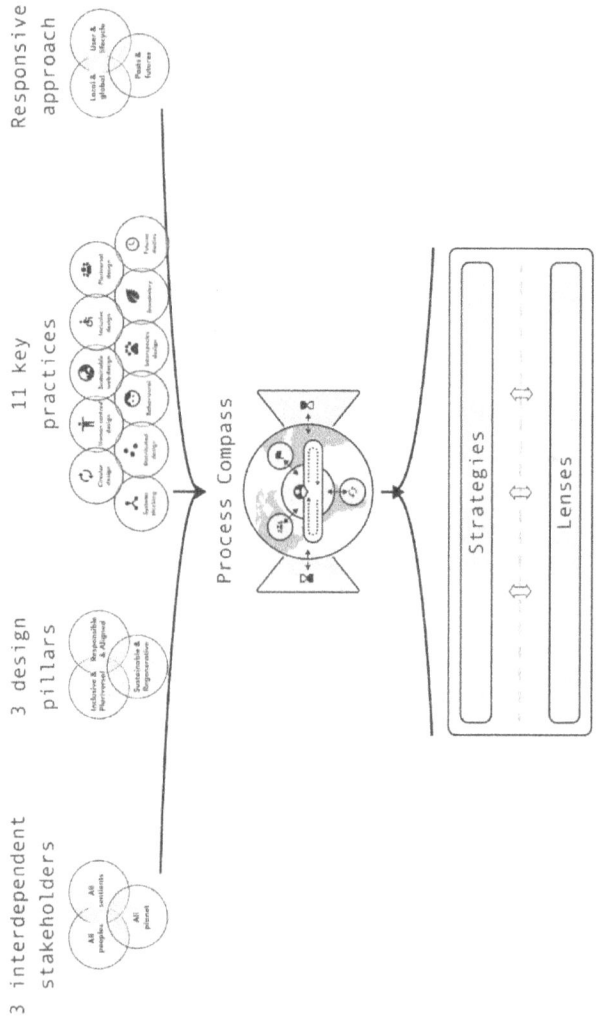

Figure 4 - Life-centred design Future Snapshot

# The Future Snapshot

Since all the commonalities and variations contributed to an alignment with the global goals and the Doughnut Economy, an optimised life-centred approach should emerge when they are all combined in a synthesis that:

- Balances consideration for environmental, human, and animal impacts
- Includes social justice and mindfulness of the power of design
- Includes digital and phygital strategies

Thinking in terms of how life-centred design differs from human-centred design, this merging generated the five key components of the Future Snapshot framework (Figure 4 - Life-centred design Future Snapshot):

- **3 Interconnected stakeholders** to protect
- **11 key supporting design practices** to work at micro, meso, and macro levels
- **3 Design Pillars** to uphold
- **A Responsive Approach** to respond to changing levels and stakeholder needs
- **The Life-centred Design Compass** to visualise how the *responsive approach* uses the *11 key design practices* to uphold the *design pillars* and protect the *interdependent stakeholders*
- **Strategies, lenses, and methods** to implement life-centred design

# 2.2
# Interdependent stakeholders

Life-centred design's stakeholders can be identified as three large groups, with human-centred design's target-user and business stakeholders remaining at the centre (but no longer considered alone):

- **All peoples**

  o Target users
  o Non-users—Individuals, communities, and employees of organisations working within the product lifecycle
  o Invisible humans—Individuals and communities not involved in the lifecycle but who are impacted by it
  o All human knowledge and ways of existing

- **All animals**

  o From large animals (amphibians, reptiles, birds, and mammals) to insects and microbes; on land, sea, air, or underground; domestic, livestock, captive, or wild; whether 'proven' sentient or not

- **All planetary resources and ecosystems:**

  o Vegetation (trees, forests, swamps, etc.)
  o Water systems (oceans, lakes, rivers, freshwater)
  o Air
  o Soil
  o Climate and weather
  o Landforms (mountains, hills, etc.)
  o Sunlight
  o Noise
  o Temperature
  o Gases and atmospheric elements
  o Biodiversity

# 2.3
# Key design practices

The varying life-centred approaches draw from different mixes of design practices, with circular design, systems thinking, biomimicry, and foresight being the most common across all.

Considering all the practices from each of the approaches contribute to an alignment with the global goals and the Doughnut Economy, an optimised life-centred approach emerges when all the practices are combined with human centred product design.

However, animals are majorly underrepresented in any life-centred approach or are included in the environment by default. But animals are neither human nor environment, giving them a unique spectrum of needs, desires., challenges, and ways of existing. To balance this, the framework needs to also include 'interspecies design'.

While not an exhaustive list of all the practices utilised in life-centred design, these 11 practices combine to form the core of the framework's purpose—regenerative design supporting a thriving planet for all life.

- **Circular design** to keep products in use, repaired, reused, remade, or recycled
- **Behavioural design** fostering sustainable and ethical user behaviour
- **Sustainable digital design** to ensure digital services (and the digital aspects of physical products) reduce their environmental impact
- **Systems thinking** to zoom out and see the lifecycle, from extraction of materials to production, shipping, use, and end of use
- **Pluriversal design** to ensure all peoples along the lifecycle are considered, and to protect the ways of existing that are different to the dominant Euro-Western way
- **Interspecies design** to consider animals as legitimate stakeholders
- **Biomimicry** to interpret nature's forms, functions, and systems into sustainable technical solutions

- **Distributed design** for distributing production to move data instead of materials to reduce waste
- **Foresight** to consider future impacts of today's decisions
- **Human-centred** and **inclusive** product design at the core

Understanding these practices will enable designers to better apply the strategies, lenses, and methods to experiment and invent their own.

Following is a short introduction to each practice with learning exercises.

# Circular design

A core practice of life-centred design is circular design, which seeks to redesign how we produce goods and services so that they fulfil their intended purpose and meet human needs in more sustainable and regenerative ways. The term 'circular economy' first appeared in 1988 in The Economics of Natural Resources[22] initiating the idea of designing out waste and designing in regeneration.

The Circular Economy can be utilised at all levels—city, district, and product—to bring more wealth, security, and well-being to people and nature by creating new relationships between users and creators, and between citizens and their environments. It is now moving into the mainstream from being a niche topic to what many companies and governments consider a priority area of research and implementation.

The Circular Economy involves changing business models, product design, and consumer behaviour by[23]:

**Designing out waste and pollution**

- It is argued that waste is not natural, that it is a man-made phenomenon—in nature, any resource produced but not used by one organism becomes food for another. Circular design reduces waste and pollution by converting waste into a resource for something else and substituting fossil and critical materials with reusable materials recovered from products already made, which also reduces waste and pollution created from extracting virgin materials

**Keeping products and materials in use**

- Keeping products in loops of reuse, repair, renew, and recycle extends the value of their materials by giving products and their parts more than one life

**Regenerating environmental and human systems**

- Creating products and services that exist as symbiotic components of other systems also means designing in ways that heal damage and further enrich both environmental and human systems, moving sustainability from 'do less harm' to 'do more good'

The Circular Economy provides a holistic view of the true lifecycle of a product's materials. In a circular model, these materials are categorised as biological and technological (Figure 5 - The Circular Economy).

# The Circular Economy

Based on the Circular Economy model (Ellen MacArthur Foundation )

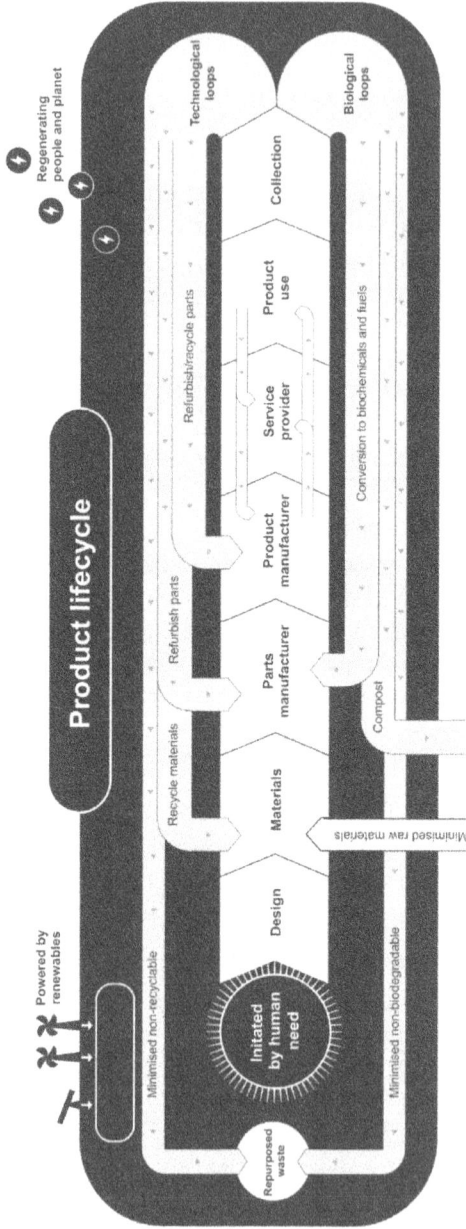

Figure 5 - The Circular Economy

## Technological Loops

On the top of the model are the technology loops showing how materials that are transformed from their natural state into new materials and product parts cycle through a circular product lifecycle.

These transformed 'technological' resources are kept as clean as possible (free from toxic chemicals or anything that with make the material hard to recycle) for reuse and recycling, and to minimise waste. The products are made as modular as possible to enable repairability and non-destructive disassembly for continued reuse, repair, remake and recycling.

- **Use, Maintain, Repair**—The customer's time of use of the product is maximised through optimised durability and ease of maintenance and repair
- **Reuse**—When the product no longer serves its initial purpose, or the user no longer needs it, the user reuses the product or its parts for other purposes, like using a bucket with a hole as a container pot for a plant, or it is passed on to someone else who can use it
- **Renew**—When the user no longer needs the product, or it is worn or damaged, it is returned to the manufacture for refurbishing and redistribution
- **Recycle**—The product or parts can be returned to the manufacturers for recycling the materials
- **Waste to resource**—Any materials not reusable by the lifecycle are converted into a resource for something else

## Biological Loops

On the bottom of the diagram are the biological loops showing how biological nutrients and matter that stay in their natural state cycle through a circular product lifecycle (paper, cork, wood, bamboo, Tencel, linen, mycelium, cotton fabrics, etc.).

Planetary resources used in their natural form are also kept clean of harmful chemicals and additives for reuse or transformation into other energy sources or returned to nature as compost.

- **Use, Maintain, Repair**—The customer's time of use of the product is maximised through optimised durability and ease of maintenance and repair of biological materials, like sewing of torn fabrics, etc.
- **Reuse**—The user can reuse parts of the product for other purposes, like natural fabrics for rags or cushion stuffing
- **Conversion to biochemicals, fuels, and compost**—Unusable waste, such as food scraps and sewerage sludge, can be treated to produce energy for

The Circular Economy in the form of biogas or returned to nature to decompose and regenerate natural resources
- **Waste to resource**—Any materials not reusable by the lifecycle are converted into a resource for something else

At every stage of the cycle, the generation of waste is minimised. Each stage can also be optimised to become resilient systems that contribute to the needs of the community they draw resources from. A life-centred business reinvests some of its prosperity into these human and natural ecosystems or into systems that reflect the business's *Life-centred purpose* (see the *Strategies, lenses, and methods* chapter).

## Key circular strategies include:

- **Take a systems view** to leverage unseen connections
- **Convert product to service**—Circular designers and businesses use a service-first mindset, seeking to minimise resource use by converting a product to a service and/or maximum digitisation of the experience. Designers and product owners work closely with manufacturers and suppliers to troubleshoot, adapt, and innovate design to maintain circular integrity
- **Extend longevity** through durable design and materials
- **Modularity** for user maintenance, upgrade, adaptability, and user-repairability
- **Dematerialise** the design so it needs as few parts as possible
- **Energy efficiency** during use
- **Safe, sustainable, and recycled materials** to optimise sustainability, durability, and resilience
- **Design for loops** of reuse/repurpose, repair, collection, disassembly, refurbish/remanufacture, and recycling
- **Design out waste,** or convert waste into a resource for something else
- Use of **renewable energies** throughout the lifecycle and supply chain
- **Digitise** with digital technology to monitor resource and energy use, monitor the circular changes, and allow customers and supply chain partners to share feedback
- **Be nature-inspired** by drawing inspiration from the sustainable and regenerative forms, functions, and systems in nature
- **Distribute and localise** manufacturing
- **Regenerate** environmental and human systems by leaving them better than how they were found

The responsibility of circular businesses to extend the life and value of materials through reuse, repair, refurbishment, and recycling also impacts the design of digital products. Digital experiences drive use of physical objects, like smart phones, cloud servers storing and handling data, the network towers, cabling, and satellites enabling the data transference, and the demand for the metals and materials to build all of these.

Digital designers can use a product lifecycle view to enable the systemic consideration of how the UX design might encourage excessive and/or unethical consumer behaviour that has a flow on effect driving discriminative behaviour and/or forcing supply chain workers into harsh and unfair labour conditions.

## Challenges

For all the benefits of the Circular Economy, however, challenges to its adoption remain. Studies divide the barriers to adoption into four categories[24]:

- **Financial**—Nature and people outside target audiences are often not considered when economic decisions are made as they are not factored into prices, a problem amplified by the GDP index not factoring in social and environmental values. There are also economic issues with the circular design models not including reference to the markets that the recycled, refurbished, resold materials compete in, making it hard to know if circular economies are making an impact25
- **Structural**—Circular design can be hampered by limited availability and quality of recycled material, lack of data, and the need for extended expert knowledge, such as system thinking and ecological principles
- **Operational**—The landscape, relationships, regulations, and short-term rewards of the current linear system make changing alliances and closing loops difficult for businesses to innovate circular design
- **Attitudinal**—Lack of awareness in business and the community, as well as cultural barriers in aligning supply chains, has kept demand for circular solutions low

There are also challenges to balancing the various circular strategies. For example, on the one hand we want to design for durability, but on the other hand we want to design for disassembly. A weld is more durable than a snap fit, but it is less easy to disassemble.

Also, there are 'rebound' and 'demand-side rebound' effects[26].

While circular design aims to slow down resource waste, closing loops are at risk of negative 'rebound' effects on the value chain and/or increasing demand for the raw materials. For example, you might improve materials with more sustainable options, and perhaps reduce ownership by converting your product to a service, but if these don't fully replace the original non-circular problems, incorporating the additional circular components might raise the overall energy consumption and waste.

Another risk is demand-side rebound, where circular changes drive an increase in purchase sales instead of fostering activities that keep materials in circular loops. If changes reduce the cost and price, this may drive uptake in new purchases. Recoverable rocketry, as an example, reduces launch material and energy needs, but making rocketry cheaper will increase the launches.

One solution is for business to offer circular alternatives in the same sales channels as their primary offerings to displace them[27].

# Exercise 1—Discover lifecycle thinking

*Tool—Lifecycle Map* (Figure 6 - Exercise 1-Lifecycle Map).
*Download from lifecentred.design*

Use the *Lifecycle Map* to map the flow of materials and parts of a product across its full lifecycle to learn about impacts on all peoples, animals, and planet.

## About the Lifecycle Map

The *Lifecycle Map* is best used after mapping your product's parts and materials to:

- Audit the flow and energy use of materials of a product to identify key opportunity areas
- Audit the environmental, social, and economic impact of a product to identify key opportunity areas
- Map an ideal future 'Ambition' state as a goal

For this exercise, however, as a starting point to initiate lifecycle thinking, just map a few of the materials and parts of a product you know. Seek data to inform the map if you like, but it's okay not to know or understand all materials and leave gaps for now.

For complete instructions for this tool, see *Strategies, methods and lenses* section.

## The Lifecycle Map layout explained

The header row is where you write your product name, tick whether it is a 'Current state' audit or an 'Ambition state'.

Also, write the key **HUMAN NEED** that the product fulfils.

The **LEGEND** suggests how to use colourised sticky notes to capture the resources, energy, impacts on people, planet, and animals, and ideas for life-centred improvements. The idea of this colour palette is that the more your life-centred strategies reduce non-renewable energy and negative impacts (red and orange sticky notes), the more green and blue your *Lifecycle Map* will become, visually cooling your map as you 'cool the planet'.

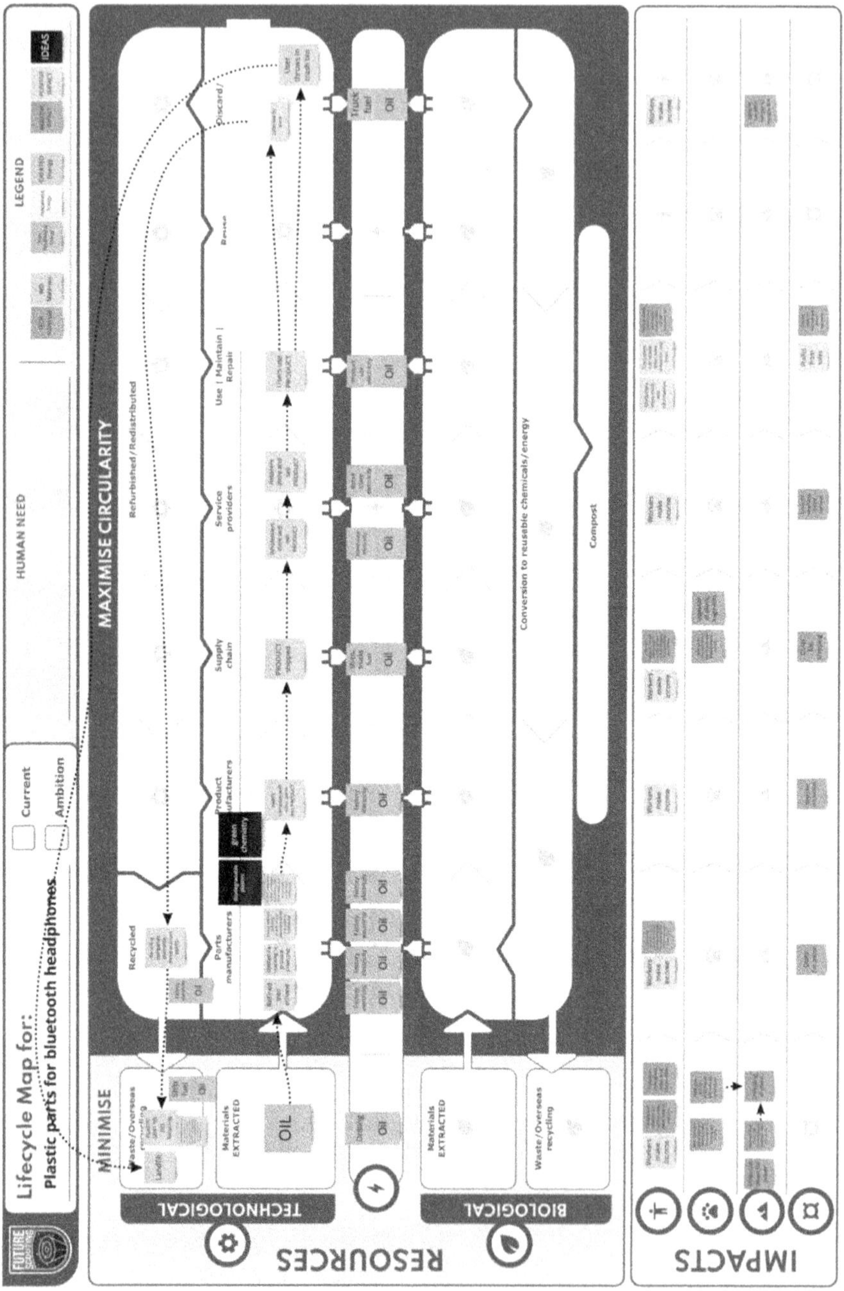

Figure 6 - Exercise 1-Lifecycle Map

The bulk of the tool is made of two sections:

- **RESOURCES**—the technological and biological resources used to create and maintain your product
- **IMPACTS**—the impacts of the product lifecycle on people, animals, planet, and finances

The **RESOURCES** section is where you capture the materials and energy used to create your product.

The flow starts in the left-hand column 'Material extraction' and flows into the **CIRCULARITY** section—this is the supply chain flow and the time the target-user uses the product. The flow then loops back into circular flows of reuse, repair, renew, recycle, or waste.

- The top resources row is the **Technological** row, to record the materials that are processed into something beyond their raw state (alloys, plastics, etc.)
- The bottom row is the **Biological** row, for materials kept in their natural state (cotton, etc.)
- The middle row is for recording the energy used/created at each step

The **IMPACTS** section is where you capture the positive and negative impacts on people, animals, planet, and finances, at every stage—these stages align with the same ones in the **RESOURCES** section above.

**Mapping the resources**

In the **RESOURCES** section, starting at the 'Materials Extraction' column, and using the materials sticky notes:

- Note down any materials of your product that you can identify, and place them in the technological or biological box
- Pick one technological and one biological material/part and map these through the entire journey, noting what you think happens to them—are they transformed into new material, attached to other parts, separated at collection or recycling, etc.?
- Next, using the energy column and the energy sticky notes:
  - Map the energy you think is used or produced by what happens to the material/part at each stage
  - Consider if any energy, heat, etc. produced isn't being used

Remember, this is just an exercise for now, it's okay not to know or understand all materials and leave gaps.

## Mapping the impacts

- In the **IMPACTS** section, starting at the left-hand side, consider the life-centred stakeholders—people, animals, and planet, and capture any impacts to them at each stage according to what happens in the **RESOURCES** section above. Consider:

  o All peoples

    - Target users
    - Non-users—Individuals, communities, and employees of organisations working within the product lifecycle
    - Invisible humans—individuals and communities not involved in the lifecycle but who are impacted by it
    - All human knowledge and ways of existing
    - *Create human personas for the key humans*

  o All animals

    - From large animals (amphibians, reptiles, birds, and mammals) to insects and microbes; on land, sea, air, or underground; domestic, livestock, captive, or wild; whether 'proven' sentient or not
    - *Create animal personas for the key animals*

  o All planet

    - Vegetation (trees, forests, swamps, etc.)
    - Water systems (oceans, lakes, rivers, freshwater)
    - Air
    - Soil
    - Climate and weather
    - Landforms (mountains, hills, etc.)
    - Sunlight
    - Noise
    - Temperature
    - Gases and atmospheric elements
    - Biodiversity
    - *Create environment personas for the key environmental elements*

  o In the bottom row, map the finances

    - Costs and profits to the business

**Assess**

You should then have a visual representation of the life cycle of your product—who and what it's connected to, and what it takes and gives back.

- How much red and orange do you see? This is an immediate visual indication of how life-centred your product is
- How many materials/parts stay inside the **CIRCULARITY** section, and how many pass outside the loops and end up as waste? Can any of these waste materials be changed to keep them within **CIRCULARITY**, or can they be converted to a resource/fuel for something else?
- Are there concentrations of negatives to focus on? How would you redesign the product to stay in loops?

Delve deep until you get more understanding of product lifecycle thinking.

# Inclusive design

I once met with someone who was blind to learn how they navigated digital experiences using screen reader technology. At times it was excruciating to watch.

Without accessibility designed into the page layout, the screen reader ranted confusing, incomplete information, and the user couldn't know everything that was on the page or how to navigate it properly. An essential task that was simple for many of us, like paying a bill online, was rendered impossible and resulted in the user receiving late payment notifications.

Inclusive design is human-centred design that considers a range of human diversity, including but not limited to age, gender identity, sexual orientation, physical ability or attributes, race, ethnicity, and belief.

Its aim to 'design for one and extend to many' inspires innovation that improves value and benefits all customers, rather than creating separate experiences that usually become unsustainable.

Accessible websites and policies, the iPhone's accessibility settings, and the electric toothbrush are all innovations designed for less common experiences which also bring new benefits to all users. Video captions, for example, not only allow the Deaf and hearing-impaired to experience videos, but they can also be extended to all users as new benefits such as for translation or for wanting to enjoy videos in noisy settings. Ramps for wheelchair access also improve access for wheeled luggage and people with walking limitations.

The Institute for Human Centred Design provides the following principles for inclusive design[28]:

- **Equitable use**—The design does not disadvantage or stigmatise any group of users
- **Flexibility in use**—The design accommodates a wide range of individual preferences and abilities
- **Simple, intuitive use**—The use of the design is easy to understand, regardless of the user's experience, knowledge, language skills, or current concentration level
- **Perceptible information**—The design communicates necessary information effectively to the user, regardless of ambient conditions or the user's sensory abilities
- **Tolerance for error**—The design minimises hazards and the adverse consequences of accidental or unintended actions
- **Low physical effort**—The design can be used efficiently and comfortably, and with minimum fatigue

- **Size and space for approach & use**—Appropriate size and space are provided for approach, reach, manipulation, and use, regardless of the user's body size, posture, or mobility

More recently, for digital experiences, Microsoft developed their Inclusive Design approach which utilises biases as opportunities for innovation to improve experiences not just for those excluded, but for many.[29]

Accessibility advocates Henny Swan, Ian Pouncey, Heydon Pickering, Léonie Watson provide this simple and succinct list of accessibility interface design principles[30]:

- Provide comparable experience
- Consider situation
- Be consistent
- Give control
- Offer choice
- Prioritise content
- Add value

And author and media scholar Böjrn Rohles shares principles for inclusive UX in his excellent article *Principles for diversity in UX design*[31]:

- Respect diversity and understand it as a strength for society and design
- Ensure functionality and comprehensibility for all user groups
- Consider ethical trade-offs in every design decision
- Keep an open mind and question all design decisions
- Ensure flexibility and customisability of the product or service
- Develop digital products and services iteratively and test them continuously with as many (and as diverse) people as possible
- Prioritise security, privacy, accessibility, and good user experience
- Consider the design from many angles (systems thinking) and weigh the effects of a design in a context for which it was not intended

## Intersectionality

Another diversity aspect coming more into focus is intersectionality—when different aspects of a person's identity expose them to overlapping discrimination and marginalisation, whether by societal attitudes, systems, or structures.

Intersectionality further expands the understanding of diversity by its inclusion of more socioeconomic factors, like mental health, criminal and medical records.

Racial and gender biases are already well documented in the perpetuation of 'sexist hiring practices, racist criminal justice procedures, predatory advertising, and the spread of false information'[32].

And just as racial bias is well documented in AI, the systemic wealth, health, opportunity, and power inequalities for the LGBTQ+ community can be perpetuated without innovation de-centred from heterosexual and cisgender focused data.

Below are two different inclusivity perspectives to consider the intersectional experience for a white, cis, non-disabled woman who identifies as lesbian:

**1. Designing for gender equality**

A synthesis of various principles lists advocating for gender equality:

- Only collect gender-specific data if it's necessary
- Use gender-neutral language
- Be upfront about intent when collecting data
- Avoid suggesting or perpetuating hierarchy by listing info/fields in alphabetical order
- Give options recognising diversity when collecting data

**2. Designing for the LGBTQ+ community**

Jason Tester, a research affiliate at the *Institute for the Future* and board member of the US National LGBTQ+ Task Force, initiated *queerthefuture.org* to foster futures of resilience and freedom from ongoing hostility. From his 17 years of experience in research and co-creative activities, Tester provides the following insights for including LGBTQ+ values[33]:

- Go beyond democratising access to innovation by identifying and dismantling inherent biases and discriminations in research and development
- Find comfort outside categorisation
- Understanding how different aspects of a person's identity can expose them to coincidental discrimination and marginalisation to foster the full richness of diversity
- Prioritise the pleasure and joy that the community have always celebrated long before they were recognised and valued for their 'differences'
- Leverage the community's adaptability and resourcefulness born from their oppression

# Pluriversal design

Pluriversal design is a social innovation practice that 'acknowledges different ways of being, acts to mitigate injustices perpetuated through design, and proposes community-based alternatives that represent the multiplicity of lived live experiences that we can achieve together[34]'. The pluriverse is a perspective of the world that recognises there are 'many centres', many ways of being that should thrive equally, not just the one dominant Euro-Western-centric view of constant growth.

Pluriversal design was born out of the Global South in response to these values being embedded in globalisation and modern design, which have slowly destroyed older and alternative knowledges and ways of being. Its methods decentre the designer from the process, allowing those being designed for to lead the process and design their own solution, ensuring their values and ways of life are embedded and protected.

For example, the Wixáritari, an Indigenous Peoples in San Miguel Huaixtita, México, recognise and plan by cycles and signs in nature, that aren't well represented on the Western calendar. This caused tensions when local harvest rituals conflicted with business calendar systems based on the Gregorian calendar which recognised only Western holiday rituals.

*Maria Rogal*, a professor of design at the University of Florida, co-designed with a group of Wixáritari teachers, community leaders, and youths to translate their concepts of time and cycles into a more locally and culturally relevant calendar by placing the Western calendar on the margin and focusing on the natural signs and 'colours of corn'.

By collaborating and localising the design process to be led by local values, the calendar also captured local stories and traditions, making it a tool for passing on community history and wisdom.

While lifecycle and systems thinking zooms the design process out to see the problem as part of an ecosystem, pluriversal design zooms into specific locations and their distinct values.

It aligns life-centred design with the people-focused SDGs and the 12 dimensions of the social foundation of the Doughnut to 'improve the lives of marginalised individuals whether within or across countries and populations.'[35]

But pluriversal design is beyond decentring whiteness, maleness, straightness, non-disabledness, etc. It's about decentring the current dominating Euro-Western-centric ways of existing:

- **Capitalistic**—Individual and private ownership of resources for the purpose of personal profit

- **Hierarchical**—The culture of order and rank according to pre-defined levels of importance
- **Patriarchal**—Male domination and privilege
- **Colonial**—The control or governing influence of a nation over a dependent country, territory, or people
- **Heterosexual and Cis**—A binary male/female concept including only cis people, who identify as the same gender as they were presumed at birth, and who are attracted to the opposite sex
- **White supremacist**—beliefs and ideas fostering or perpetuating a sense of superiority of the lighter-skinned, or 'white' human races

It is also about challenging perspectives and terminology, such as 'inclusion' itself, which suggests 'others' were never in the process and need to be invited in[36].

Pluriversal design takes us deeper into the philosophy of design, by recognising that what we design affects our ways of existing, and vice versa, by the 'rituals, ways of doing, and modes of being' generated by engaging with what we design[37]. If design solutions don't allow for a variety of ways of existing, according to the needs, problems, desires, and resources of various locations and cultures, these designs can end up conforming, excluding, repressing, or erasing those cultures, their languages, and wisdom.

## A world where many worlds fit

In his 2017 book, *Designs for the Pluriverse*, Columbian American Anthropologist and distinguished critic of development, Arturo Escobar, proposed the concept of the Pluriverse as a 'world where many worlds fit', in contrast to the 'universal' single world perspective created by modernity[38].

Modernity refers to the dominating capitalistic, political, and economic values that shape the society, economy, and ways of being in the Global North. It champions scientific rationalism over tradition and myth, human mastery of nature, individuality, continued growth and progress, and dependence on governmental institutions[39].

'A world where many worlds fit' was first stated by the Zapatistas, an anti-globalisation social movement in Mexico seeking First Nations Peoples' control over local resources and land. Escobar formalised this thinking into the pluriversal framework to address the diversity of worlds suppressed by colonisation and its focus on development.

At the frameworks' core are two attributes:

- Not all countries must evolve according to the Euro-Western capitalist mantra of development, including economic growth and material progress

- African, Asian, and Latin American alternatives that incorporate non-Western concepts of what constitutes a thriving society should be elevated in discussion to expand ideas of possible futures

## Pluriversal design and the SDGs

While respecting the many positive elements of the SDGs, however, a cry from the Global South argues that innovating purely through technical and managerial solutions, without socio-cultural transformation, will not lead us out of our wicked problems. This is echoed by the Doughnut's argument that human commerce cannot be separated from the human condition.

Authors of *Pluriverse: A post-development dictionary*, a collection of 100 essays on alternatives to globalisation, argue the capitalism of the 1980s distorted the sustainability focus of earlier incarnations of the SDGs. Since then, they argue, the SDGs have become 'a programme of poverty alleviation' without delving deep enough into causes of poverty and have become a movement to solve the 'development problems' of the Global South by funding them into becoming capitalistic replica states of the Global North[40].

As an alternative, the authors suggest the following should be the values of sustainable and just societies that pluriversal design can foster:

- Diversity and plurality
- Autonomy and self-reliance
- Solidarity and reciprocity
- Commons and collective ethics
- Oneness with the rights of nature
- Interdependence
- Simplicity and 'enoughness'
- Inclusiveness and dignity
- Justice and equity
- Non-hierarchy
- Dignity of labour
- Rights and responsibilities
- Ecological sustainability
- Non-violence and peace

While the UN have, at time of writing, recently published a call for perspectives and inputs to the Global Sustainable Development Report 2023 from a wide range of diverse perspectives and disciplinary scientific backgrounds[41], a world of true diversity and inclusion should enable thriving states that don't have continued growth and technology solutions as their key focus[42].

This is important to know right now as many design projects and businesses—including instances of life-centred design—are aligning themselves with the SDGs. While having these common global goals are helpful in so many ways, without ensuring plurality, these goals run the risk of imposing blanketing solutions that became popular with designers due to successes that align only with Global North values and imposed by trying to teach communities how to 'improve' according to values that aren't theirs.

Pluriversal design elevates quieter voices to probe deeper into the root causes of the problems that global goals sincerely aims to address. It doesn't directly aim to dismantle the dominant 'One World' view but to multiply it into many co-existing centres and expand inclusion to those who are often excluded or misrepresented in main design narratives. But it does also raise the question if modernity's disrespect for environmental limits, indirect fostering of power monopolies that erode stability, and marginalisation of Indigenous knowledge and ways of being can exist within the sustainable, regenerative, and just pluriverse.

## Alternate pasts

By knowing and acknowledging multiple perspectives of experience and being, pluriversal design also challenges the modern idea that we all live in a single world with a single past.

The unique food gathering methods of Indigenous Australians, that ensure enough food and resources are available for all the community season after season, are some of the most sustainable in history[43]. This management of resources allowed Indigenous Australians to survive for thousands of years, yet such knowledge is not mentioned in the most accessible internet-based resources. This systemic lack of plurality perpetuates itself by making the pluriverse invisible to designers of privilege.

Recognising past wisdom allows it to be retained to form a hybrid of modern and traditional knowledge. Pluriversal design reminds us of this connection between people and their place, their past, and their future.

## Redesigning design education

Pluriversal design addresses not only the practice of design but also design education. It recognises that since design education is an elite practice taught in the Euro-Western world, it favours the values of modernity, which restricts many designers' ability to see and design beyond their single world view.

For example, the cultural richness and wisdom embedded in languages can often be lost when participants or students are forced to participate in one dominant language. The preferred language of English in many studies demands students depart their native tongue before they can learn these skills. But while the English language uses words like 'things' and 'its' to differentiate humans from animal entities, most Indigenous languages do not because their cultures consider all life as equal[44]—this disparity in language can distort or repress the continued flourishing of diverse cultures and interpretations of the world.

## Implementing pluriversal design

Pluriversal design deconstructs the design process by decentring the designer. They act as a facilitator who lends their skills and self to enable the people to design the solution for themselves and with their values.

Pluriversal designers collaborate with local people and their values, knowledge, resources, and cultural assets to mitigate injustice perpetuated through modern design. In this way, pluriversal design decolonises design and allows for solutions that are made with, and for the people's unique ways of being.

In his talks and workshops on plurality, recommoning, and designing for alternative economies, Washington-based designer, researcher and educator, Dr Dimeji Onafuwa highlights the need for understanding that there are many perspectives of 'better' when design attempts to improve experiences—without accounting for plurality, we design ourselves into one future where not all can thrive. He also notes the distinction between representing different needs in a design solution and representing different *worlds*.

Inspired by a conversation with Escobar about transitioning from the idea of single ownership of things to a collective ownership of common resources, Dr. Onafuwa created a set of guiding principles for pluriversal approaches to UX design to be used 'humbly, collaboratively, and systemically'[45]:

- Be egalitarian—engage stakeholders and diverse cultural understandings early and often
- Fight symbolic violence
- Act as an ally—by going into others' world experience
- Be multi-modal—zoom in and out of system views to work across scale, scope and time
- Adopt collective agency—view all angles of the problem and consider animal stakeholders
- Steward the rights of others
- Lend your privilege upstream
- Stay with the problem—apply a long-term posture and mindset

When asked if there might be a priority of these, Dr Onafuwa suggested there is no hierarchy of these principles, that they are 'a trojan horse to induce discussion', but that designers could start with 'Stay with the problem'.

Rogal, who has 15+ years of experience experimenting with pluriversal approaches, recommends:

- First, understand design's role in communication and the creation of culture and power
- Work in context early
- Dialogue, participatory research, and creation are key to fostering equality and decolonising concepts
- Use various forms of methods and mediums—spoken, visual, artistic, ritual, etc.—to foster participatory research and creation.
- By decentring Euro-Western-centric design approaches, 'we make design available and vulnerable in the face of knowledges and practices that are central to those with whom we work'[46]

Embracing the complexity of a world of many centres also means gaining access to the broader spectrum of opportunities for exciting innovation inspired by people who have defiantly lived their true values in worlds not designed for them by hacking the very systems and tech that excluded them. Some communities, the LGBTQ+ for example, survive and seek to thrive in both Global North and Global South by hacking systems and technologies and developing underground networks to express their true selves and to live according to their values—and sometimes just to survive[47].

There is much value in the skills and knowledge born from the struggle against repression.

To filter pluriversal thinking down into micro-level design, we can draw from pluriversal techniques as they are implemented in social innovation.

## Case studies

### Critical Tools: Design pedagogy for alternative ways of making futures
Deepa Butoliya

The PIVOT 2021 conference, themed as *Dismantling/Reassembling*, explored tools for 'dismantling normalised ways of thinking, being, and making', so that a community's resources could be reassembled in new ways to foster sustainable and just futures[48].

After experiencing floods in her home in Detroit, US, Butoliya discovered the floating amphibious vegetable plots in the flood-prone areas in Thailand that were created using local wisdom and making. *Jugaad* is a Hindi word representing 'making do' in situations of lack of resources, often employed in experiences of oppression, and therefore represents a critical making practice in the Global South.

Drawing on *Jugaad's* inherent resilience and minimalism to infuse more sustainable thinking in design education, Butoliya set her students a task to choose a 'blackbox' item from their personal environment—something of which they had little understanding about what went on inside to make it work. She challenged them to dismantle it and reflect on who and what was impacted by all its parts, and the lifecycle of all these parts. Students were then asked to reassemble an alternative item with found objects from their immediate environment.

Butoliya's work recommends the following:

- Disassemble to reassemble to challenge the immediacy and materialism of modern design and the value of its default sleek and manufactured styling
- Challenge colonial and Euro-Western-centric foundations in industrial design to seek more sustainable materials and methods
- Understand and utilise the value of knowledge born from the struggle against repression and the survival experiences of the marginalised and Global South
- Add a future lens to the act of repair and repurposing to explore alternative ways to design that foster more sustainable and just design

### *Rolling Stones: Dismantling, reassembling, and reimagining possible tools through collaborative story-making approaches*

*Manuela Taboada & Dr Jane Turner*

In this presentation from PIVOT 2021, Taboada and Turner addressed the challenge of staying in the Pluriverse mindset when working within existing modern capitalist design systems[49].

To address this, they adapted the analogue RPG gaming process into a design approach that fostered active imagination, to protect it and the process from being drawn back into non-plural thinking and ways of designing.

They suggest varying workshop tools, from dice and cards to sticks and string, and challenging participants to respond to 'twists' in history. For example, in a project exploring a world without plastic, they explored speculative scenarios where future 'agents' would visit homes to collect banned items of plastic.

Their goals are to facilitate imagination rather than find solutions, and for the process itself to be the sharing of worldviews, as a means of dismantling standard design thinking and process.

- Stimulate dialogue and interaction over answers
- Gamification
- Focus on process over objective
- Focus on creating over telling and listening
- Allow for change, in process, rules, etc.
- Allow multiple worlds and narratives to appear
- Foster emergence rather than focus on utopias

For Taboada and Turner, allowing participants to become world- and story-builders allows emergence/evolution in the process and creates alternative ways for others to encounter multiple worlds—all of which that fosters active imagination.

### *Decolonizing Our Future Through Inclusive Storytelling*
*Pupul Bisht*

Bisht, a futurist and designer from India, argued in her presentation *Decolonizing Our Future Through Inclusive Storytelling* 'to authentically talk about alternate futures, we need to talk about alternate histories'.

Bisht highlighted the importance of recognising alternate histories of local knowledge and experience of those we design for and with—not just past wisdom, but past experiences and perspectives to decolonise the past by recognising we don't have a singular history[50].

### *The Tin Can Radio*
*Victor Papanek*

In *Design For the Real World,* Papanek shared a wonderful story from 1962 about designing a radio for illiterate people in villages who were unaware that they were part of a nation-state.

The radio was made from a discarded tin can and used the heat converted from a wax candle inside made from local resources, like paraffin or cow dung, and heated by thermocouples, to activate an earplug speaker to broadcast the national news—costing less than 9c per radio.

Papanek deliberately excluded aesthetics to respect the many different cultures utilising the radio, which allowed regions to customise it to their own culture.

*****

At its most simple, pluriversal design is a practice that is 'attuned to justice and the Earth[51]'. This takes us back to the two most important relationships recognised by many First Nations Peoples—the relationship between people, and that between people and land[52], with the relationships between people always contingent on the relationship between people and land, for stability and security[53].

If the damage to both human diversity and biodiversity have arisen out of the destructive impacts of a dominant focus on globalised development, then perhaps the path to restore the environment's health is through re-nourishing humanity's pluriverse of differences, wherein lies the many ways of knowing how to live more connected with the earth, and to heal it[54].

# Exercise 2—Know your privilege and power

*Tool–Powel Pixel* (Figure 7 - Exercise 2-Power Pixel).

*Download from lifecentred.design*

Designers need to be aware of their power and the potential impacts their work can have by understanding the power of design, how it's used, and by whom—from deciding who and what knowledge beliefs are included or excluded in research, design, and decision making, to how design influences the sourcing and production of the materials that impact different places and peoples.

The *Power Pixel* is a kind of mindfulness token for designers and individuals to hold in their hand to keep the power of design high in their consciousness and to remember their place in context.

## About the pixel

There are four sides of the Pixel to complete:

- I will lend and call out PRIVILEGES
- I will look through EXPERIENCE LENSES
- I will champion MY VALUES
- I will challenge MY ASSUMPTIONS

### Determine your Values

Core values are our highest priorities and our deepest beliefs. Once you know your values, you can use them to guide your behaviours, decisions, and actions to ensure you are creating solutions that will foster the future you want

- Think of three good unforgettable experiences, choose value words to describe them
- Think of three unforgettably bad experiences, choose value words to describe them, then flip them to their opposite
- Think of three people you admire, choose core value words to describe them
- Think of three people who really challenge your values, choose value words to describe them, then flip them to their opposite
- Think about the things in your life and name the top three you couldn't live without. Associate the feeling of them with a core value.
- Group all the similar values to identify the three most important to you.

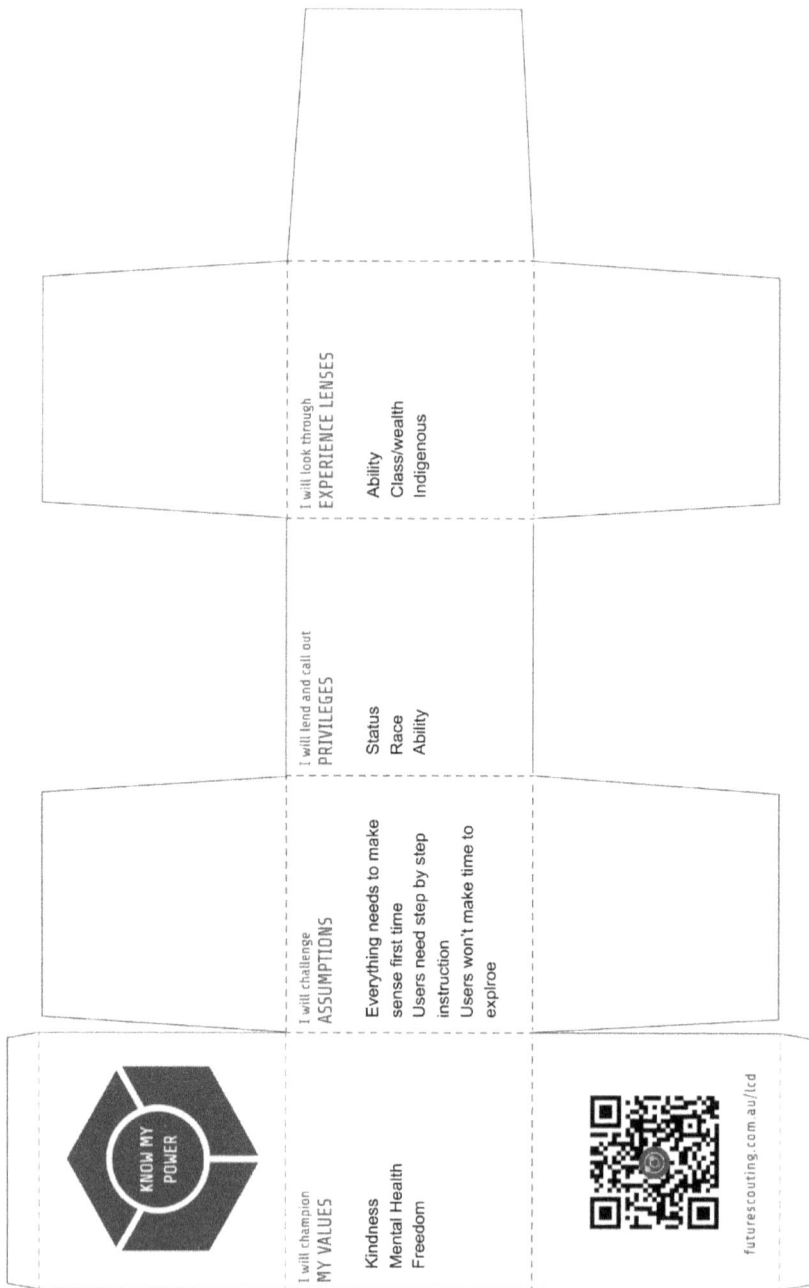

Figure 7 - Exercise 2-Power Pixel

I will champion
MY VALUES

Kindness
Mental Health
Freedom

KNOW MY POWER

I will challenge
ASSUMPTIONS

Everything needs to make
sense first time
Users need step by step
instruction
Users won't make time to
explroe

I will lend and call out
PRIVILEGES

Status
Race
Ability

I will look through
EXPERIENCE LENSES

Ability
Class/wealth
Indigenous

futurescouting.com.au/lcd

Other methods to determine your core values:

- Values discovery activities by CEO Sage—
  *https://scottjeffrey.com/personal-core-values/*
- Core value worksheets by Tommi LLama—
  *https://tomillama.com/personal-core-values/*

**Determine your Privileges**

Privileges are the unearned, identity-based benefits we inherit just by being part of one group, that we can often be unaware of, and are enjoyed at the expense of others. Knowing them allows us to use them to help the less privileged, or to denounce them to rebalance inequity

The Privilege Web* in the *Power Pixel* template guide captures key privileges and allows you to mark yourself against each.

- Map the size of your privilege:

  o Go around the web and mark where your privilege is for each segment
  o Use the blank segments to add any other privileges you think should be included (if you leave them blank, mark yourself at the highest level/outer ring for these)
  o Draw lines between each mark on the privilege segments, and colour in the emerging shape with a highlighter (as per the example) to reveal the size of your privilege

- Understand the impacts of your privileges

  o In the first outer ring, 'PRIVILEGE BENEFITS', use green sticky notes to capture how your privileges benefit you
  o In the inner circle, 'OPRESSION EFFECTS', use red sticky notes to capture how your lack of privilege affects you, and how your privileges affect others

- Think of ways to mitigate the negative effects

  o In the outer-most circle, 'MY POWER', brainstorm how to use your benefits as powers to:

    ▪ Help those less privileged
    ▪ Help the more privileged see what they have to lend

- Consider which segments most relate to your values
- Choose a few ideas and write these on the **I will lend and call out PRIVILEGES** side of the Pixel

You could also add the people most affected by your privileges to your **EXPERIENCE LENSES**.

*The Privilege Web was based on the great power literacy work of Maya Goodwill in her *Field Guide to Power Literacy*. I highly recommend using the *Field Guide's* exercises to build up 'your awareness of, sensitivity to, and understanding the impact of power and systemic oppression in participatory processes[55].'

### Determine your Experience Lenses

The Experience Lenses represent the human experiences unlike your own, to see the design experience through their eyes, to lend your privilege as a positive power, and to ultimately design solutions that celebrate the full spectrum of human experience and ways of being.

- Consider how your product might impact people who:
  - Experience life differently from you
  - Are often not included, misrepresented, or marginalised by mass design
  - You may have bias against
  - You just want to have more empathy for

- Think in terms of:
  - Accessibility
  - Gender identity
  - Sexual orientation
  - Culture
  - Age
  - Body type
  - Personal belief
  - Pluriversal experiences

- Consider which segments most relate to your values—for example, if one of your core values is mental health, you could include the consideration of users on different mental health spectrums
- Consider which segments most relate to the SDGs identified in your *Life-centred Purpose* tool

### Determine your Assumptions

We all have established opinions that generate automatic responses affecting our thinking and behaviours. These assumptions can negatively influence designer's choices of solutions and participants.

For the *Pixel's* purpose, understand your personal assumptions specific to the theme of the project's purpose so that you can challenge the assumptions you default to.

- Consider:
    - When you think of the theme, what do you automatically assume about how the solution might need to be—style, materials, order of experience phases, etc.?
    - What do you assume users and stakeholders think and need?
    - What do you assume can't be done?

Group your assumptions into a few key insights and write these on the assumptions side of the Pixel.

## How to use the Pixel

Once completed, print out your Pixel, fold it into its cubic shape (no adhesive required), keep it on your desk during the project, and define your ritual of use:

- At the start of each day
- Before any meetings, design sessions, or research sessions
- At times of decision making, confusion, or frustration
- Any time you feel you need to tune into your place in context

By its tangible nature, the *Pixel* prompts designers to pause and tune into the power of their privileges, values, assumptions, and biases as they design to help create kinder and more mindful experiences for all humans, animals, and planet.

Our core values remain the same over longer periods, but biases and assumptions can change with each project, so you might want to create new *Pixels* over time or per project.

# Systems Thinking

For the Indigenous Peoples of Canada, Australia, and New Zealand traditional 'health care' does not involve the Western approach of putting the patient at the centre of the treatment, but rather sees them as part of a 'web of connections' between the individual and their environment. The dynamic relationships between the individual, world, and culture become the focus, providing significant health benefits[56].

Maintaining this 'relationality'—the awareness of individual dynamic relationships with the environment and other lifeforms, and the strengths and weaknesses of these connections—is at the foundation of these Indigenous Peoples' ongoing wellbeing[57].

The awareness, respect, and inclusion of the connections, flows, and relationships between people and environmental systems is also the purpose of systems thinking.

Everything is connected—humans, animals, plants, machines, the weather, the economy—and this deep and complex interconnectedness allows for single leverage points to have compounding effects.

To put simply, systems thinking is a way of looking at and talking about the world as whole, not as individual parts, but by understanding relationships and patterns to identify points of leverage[58].

Donella Meadows, American environmental scientist and author of *Thinking in Systems: A Primer* in 1993, introduced leverage points as points in a system where small change can create big impact[59].

Systems thinking is a mindset and a set of tools to holistically view the complexity of our world as interconnected and interdependent systems, to see beyond the apparent and linear views, to identify the root causes. It is opposite to the current dominant view that sees things as many separate and independent parts—a view that can oversimplify perspectives and see parts out of context, and sabotage attempts to improve a system to actually make them worse[60].

System thinking can be applied to global, state, city, community, and organisational levels, or to a specific problem domain, and allows us to investigate:

- What is evolving rather than just what is planned
- The whole relationships rather than just the parts
- How components exchange and transform energy, information, and materials, and their cause and effect
- All the output, from desired outcomes to the unplanned[61]
- Understand how the system works overtime, exposing latent momentum that can create change not initially recognised

- Make clarity of the complexity by identifying patterns, routines, relationships, resources, power structures, and values
- Get to the root cause of problems
- Identify non-functioning patterns and leverage points of change, innovation, and adaptability

Shifting from our linear economy to a circular one requires systemic solutions. Considering the impacts of the true lifecycle of a product, and how everything is connected, takes designers into systemic thinking[62]. Circular design uses systems thinking to look beyond the product or service, and to see the connections of our creations and processes along the entire lifecycle, so we may intervene in ways that optimise the entire system[63].

Systems thinking, however, can be messy, complicated, and overwhelming, which risks over-analysis, looking too broadly, and losing focus. Techniques and approaches are often very technical and can require the help of specialists. Its non-linear perspective challenges the dominant reductionist thinking as taught for decades, and therefore it can be hard for designers and fast-paced, profit-driven design projects to adapt to[64]. And systemic interventions often don't evolve due to the time they take when focuses change.

But the ability to map holistic views and relationships is more important now with current unprecedented challenges making our wicked problems larger and more complex. These unpredictable times thrust us into a fundamental shift where linear thinking alone no longer works. It could be argued it never did, that it just delayed the negative consequences coming back amplified today.

Product Designer and MHCI+D alumni, Tyler La, suggests designers should practice becoming comfortable and confident with uncertainty by defining a certain level of uncertainty as acceptable to move forward[65].

## Systems thinking in design

Systemic thinking is far from new, with its colonial roots in Jay W. Forrester's approach to understanding social and engineering systems (*Systems Dynamics*, 1956) and Karl Ludwig von Bertalanffy's *General Systems Theory* (1968). This thinking evolved through the 1990s to include more environmentally sensitive theories such as Industrial Ecology, from which today's systemic design evolved.

Meadow's *Thinking in Systems: A Primer*, circulated informally due to her sudden death in 2001, until it was edited and published by Diana Wright in 2008. The book was fundamental in bringing systems thinking out of the realm of science and into the world of innovation and finding practical solutions from complexity.

Systems thinking eventually became incorporated into design thinking to evolve into today's contemporary human-centred, systems-oriented design approach. With growing awareness of the impacts of design at all levels, systems thinking has become a more prevalent discussion point as it can inform social and sustainability improvements and help businesses move from a product and ownership mindset to a service one.

In early 2021, the Design Council published their *Beyond Net Zero: A Systemic Design Approach*, where they have embedded systems thinking as an integral part to their approach, connecting stakeholders across all levels, and often, throughout the design process[66].

Using its sensitivity to the circular nature of the world, systems thinking can take design beyond service design blueprints and ecosystem maps by reframing them to uncover interconnected relationships. Incorporating systems thinking at the micro-level is about developing the ability to shift between human-centred design and life-centred design dynamically throughout the design process.

From a life-centred design perspective, system thinking can help identify where modifications can be made to the product or ecosystem for efficiency and to 'dematerialise' it:

- How might we achieve the design with a smaller number of elements?
- How might we reuse or combine elements in the system serving a similar purpose?
- How might we reorganise the relationships to optimise the system's structure?

Creating visual system maps also help reveal connections of value to broader environmental and social issues, so regenerative solutions can be designed to nourish those values, and, in this way, connect the design project to global goals.

To map and affect systems optimally, it is important to understand how to determine them.

## What is a system?

In *Thinking in Systems: A Primer*, Meadows defined a system as 'a set of elements or parts that is coherently organised and interconnected in a pattern or structure that produces a characteristic set of behaviours, often classified as its function or purpose', with three essential components being elements, interconnections, and function/purpose[67].

Systems and their purposes emerge from the synergy of their interdependent parts forming a repeatable and recognisable pattern of a combined function or purpose, like a local community, a forest, or an organ[68].

The purpose defines the boundary of the system, which we can understand by starting with the elements:

- **Essential elements**—Also called nodes, elements are representations that describe who, what, where, and when. They can be tangible or intangible, social, environmental, economic, and technological
- **Interconnections**—Interconnections provide the flow of resources and/or information (the *stocks* of a system). These can be mapped by drawing arrows between elements to show the *flow* of the stocks. Stocks are measurable resources that increase or decrease over time (materials, information, etc.), and flows are the *rates* of those changes over time.
- **Purpose**—While acting independently, with independent goals, it is the sum of the elements' interdependent relationships that forms the system and defines its emergent purpose. The purpose helps define the system's boundary. Which elements contribute to the system's purpose? If you don't include an element, and the system can no longer properly fulfil its purpose, then that element is essential. The system may contain subsystems and be a part of a supersystem or larger environment of other systems. The boundary might have filters to maintain integrity and autonomy, like an IT system firewall or a body's immune system.

## System Mapping

Systems Mapping is a powerful visual modelling tool to capture the system's elements and interconnections to create a shared vision, foster collective intelligence, and help identify leverage points for change. System maps are often used at a discovery phase, or for big problems where you don't know where to start. They can also be used at any phase of design to zoom out and diagnose a part of the user journey, supply chain, end-of-use phase, etc.

Like foresight, systems maps are not designed to predict what will happen—they enable exploration of what could happen when different factors or changes are applied. It's also good to remember that systems maps are just artificial representations of reality, so they can have gaps that can prevent us from seeing all the details.

Not all design challenges will need a systems approach, particularly when a challenge is self-contained, and well understood. But it can inform micro-level design by zooming the process out to view the interconnectedness between other elements within the project, and the projects connections to global goals, connecting user-centric design with greater environmental and social needs. System mapping can also complement a product lifecycle view.

There are various methods to map the dynamics of a system. These are technical but show where problems and opportunities lie, and allow for testing with interventional system tools:

- **Causal Loop diagrams** show feedback loops, which are circular chains of cause and effect that include multiple components, where the flow of stocks leads back to its original causing element. Feedback loops are indicated as either positive (reinforcing) or negative (balancing)
- **Stock and flow diagrams** add more detail to Causal Loop diagrams, typically through computer simulation, for analysing more quantitatively.
- **Behaviour over time graphs** can be modelled from Stock & Flow diagrams to understand greater effects of a system over time to factor in time delays of cause and effect
- If you really understand data nodes and their values, you can also use **Connected Circles Maps**

## Process

In design, a system map represents all the different elements involved in the provision of a product or service delivery, and their flows of 'stocks'—materials, energy, information, money, documents, etc. This clarifies how all the different components exchange value.

A general process for systemic design is:

- Define the problem
- Map the system
- Analyse and look for interventions
- Strategize
- Design intervention
- Test and learn

While systems dynamics is more for understanding complex problems, it can be applied to product ecosystems and supply chains, and it is a skill which 21st Century designers will need more and more.

See the *Regeneration Lens.*

# Interspecies design

While there has been much development in design thinking to include and protect people and the environment, tools to include animals in design are most lacking. Animal personhood and nonhuman and animal rights are evolving on the legal front, but how do we designers ensure our work protects their ways of being?

To champion multi-species ways of living on the planet, interspecies design is an emerging practice that combines the emerging practices of Animal Computer Interaction design, Animal Architecture, and Posthumanism to give proper consideration to animals.

The term 'animals' in this case might include creatures of varying scales, from large animals (amphibians, reptiles, birds, and mammals) to insects and microbes; on land, sea, air, or underground; domestic, livestock, captive, or wild; and whether 'proven' sentient or not.

There has been much research into interspecies behaviour by great thinkers, designers, and scholars, such as Steve North, Hanna Wirman, Anne Galloway, Patricia Pons, Donna Haraway, and others discussed below. Their research has fed into the evolution of animal technology, including technology designed specifically for animal interaction, like large buttons for dogs to open doors[69], and Pettech products like PetChtaz, a video product with a paw interface allowing pets to 'call' their owners.

While interspecies design began with animal-computer interaction, two streams of work appear to be significant to consider (although there is much overlap):

- Protect—Designing solutions for human problems that consider and protect animal ways of living
- Engage—Design solutions to be used by animals, whether via human-led or animal-initiated experiences, and for specie-to-specie and cross-species interaction

**Designing to protect**

Two examples of practical solutioning from design consideration for animals:

- Window material to prevent bird collisions—The windows in buildings caused 599 million bird deaths in 2015 and continue to contribute to their decline[70]. Students at the Sustainable Design School in Nice utilised the unique optical attributes of a few species of birds to invent a window material with properties only the birds would notice to help them avoid colliding into them
- Biodegradable and edible can ring holders—Plastic ring holders for drink cans have created perilous dangers for wildlife, particular sea animals

getting entangled in them with fatal consequences. Saltwater Brewery, a craft beer brand based in Florida, worked with We Believers, an ad company based in NY, to produce biodegradable six-pack rings from the by-products of the beer brewing process, such as barley and wheat, meaning animals can eat them and benefit from their nutrients

Design fictions, such as animal experience simulators, have explored helping humans empathise with animal experiences:

- Animal Superpowers by Chris Woebken are wearables for kids to experience animal abilities like magnifying vision to see like an ant[71]
- Interface masks by Jose Chavarria swap a human sense with animal senses like echolocation, Infrared Sensing, and Geomagnetoception to experience how animals perceive the world[72]

Senior Lecturer, Researcher and 'maker of experiences and oddities for Human and Non-Human Animals', Alan Hook designs prototypes with animals to propagate interspecies empathy, with a focus on utilising animals' love for play.

## The Interspecies Toolkit

Hook also worked on behalf of Imagination Lancaster and with Microsoft to develop the Interspecies Design Toolkit as a Speculative Design proposal complementing the Microsoft Inclusive Design Toolkit. The toolkit helps structure this emerging practice by defining various areas for consideration[73].

### Design principles

- Recognise exclusion
- Learn from other species
- Design with one, speculate for many

### The species spectrum

- Human
- Domestic
- Livestock
- Captive
- Wild

### Human engager personas

Below is an elaboration of the toolkit's human personas that represent human-animal engagement:

- Farmer
- Animal Welfare
- Consumer

- Captive parent
- Pet parent
- Hunter

**Types of Animal exclusion**

- Physical—the relationship between physical traits and abilities (arms, legs, etc.) and interaction methods (using hands versus using beak) and the physicality of human designs (product, service, architectural, etc.)
- Cognitive—comprehension and communication
- Social—the presence of animals being recognised and respected by humans

## Observe and empathise

To empathise with animals, the Interspecies Design toolkit suggests observing animals to learn about their interaction needs to prompt innovation.

You could warm up first by role-playing with friends as an animal to get familiar with non-verbal communication. Or if you have a pet or legal access to safe interaction with an animal, practice empathising with animals by allowing the animal to initiate and lead play, following its lead, to explore how that interaction is different from the usual experience of human-led interaction.

## Observe to identify[74]

Similarities and mismatches with human behaviours can then be used as prompts for innovation by exploring:

- Trust factors between humans and animals
- How animals complete the same or similar tasks to humans
- Barriers to animal accessibility to engage/interact with humans and the human world
- The animal's types of interaction, and which are most common (noise, licking, touching, brushing against others, etc.)
- How animals sense and navigate environments, and how these abilities change with different environments
- What elicits different types of responses

## Design to engage

While animals have interacted with technology since the sixties, Animal-Computer Interaction (ACI) was first introduced in 2011 by *Clara Mancini*.

Animal-computer interaction design explores how human technology affects animal experiences with the aim to improve their welfare, support their activities, and foster interspecies relationships. This is to inform interspecies design as well as research ethics, such as using kind and safe wearables to collect data on animals.

*Exploration into technology* for animals has already produced interaction innovations such as nose plate interfaces, biteable pulleys, paw activated buttons, and haptic vests, but this research was mostly based on human-initiated interactions75.

In 2019, Ilyena Hirskyj-Douglas (a Lecturer/Assistant Professor in Animal-Computer Interaction in Scotland) and Andrés Lucero (an Associate Professor of Interaction Design in Finland) delved into animal-led interactions. They explored interconnectivity for dogs with their *dog-to-dog internet project*. Designing Technology for Dog-to-Dog Interaction progressed their earlier work on dog-technology within the home to explore animal entertainment and well-being for scenarios such as enhancing the lives of pets left at home.

**Types of ACI systems to employ[76]**

- Screen and Tracking Systems
- Haptic and Wearable Systems
- Tangible and Physical Systems

**Types of play to leverage[77]**

Behavioural research has identified three main types of animal play:

- Social
- Locomotor-rotational
- Inanimate object play

Since we humans aren't very attuned to comprehending animal responses to understand the true impacts of our actions and behaviours, please be mindful of this in any experimentation, and consider engaging experts[78] from animal science, animal welfare, animal behaviour, and/or ACI research.

# Distributed Design

When supply chains were disrupted by COVID19, distributed design was embraced to globally distribute designs for local production of protective equipment (PPE) for those who needed it fast[79].

Decentralising and localising manufacturing uses global interdependent collaboration to reduce costs and waste, keep needs and systems close to users so they remain in symbiosis, and reduces negative impacts to wildlife, the environment, and invisible humans. It also supports social justice by decentralising power to enable local creation and equitable sharing of value, which in turn fosters autonomy and ongoing localised optimisation.

Distributed design emerged from the mixing of two global trends— the digitisation of design and the maker movement.

Through the open sharing of design and manufacturing knowledge, the maker movement fosters an individual's ability to be both maker and consumer. Maker spaces give anyone access to the place, tools, and resources to design and create almost anything. This democratising and decentralising of both design and manufacturing enables anyone to design and fabricate products on their own, or to collaborate with global networks of designers and makers for assistance and co-creation.

Distributed businesses use maker spaces as their nodes.

Maker spaces are generally inside a school, library, or separate public or private facility. Every maker space is unique with tools and equipment, ranging from laser cutting, soldering, wood working, and sewing to electronics, robotics, circuitry, coding, and 3D printing.

Fab Labs (Fabrication laboratories) are a particular type of makerspace that centre around digital fabrication tools (3D modelling software, 3D printing machines, etc.). With 2000+ labs in over 120 countries, Fab Labs provide aid and education about running a lab with operational support, finance, and logistics. They deploy all the materials and hardware, from laser cutters and 3D printers to large mills for making furniture and housing, to 'chapter operators' who want to set up a Fab Lab in their area. Fab Labs need to be self-funded or part of a larger project or organisation. Chapter operators are responsible for safety, operations, and contributing to collective knowledge.

Both Fab Labs and maker spaces are hubs that encourage innovation and learning through design, prototyping, collaboration, creation, repair, and fabrication. They foster community and shared knowledge and skills by attracting a mix of artists, engineers, inventors, fabricators, students, and hobbyists.

This connecting of global design knowledge with local production capabilities enables sustainability and circularity by reducing the energy and material waste from mass production and their supply chains.

The key principles of distributed design are:

- Open source
- Participatory
- Collaborative
- Eco-systemic
- Sustainable
- Regenerative
- Inclusive

Distributed design creates great opportunities for innovation by:

- Opening the design and production process to many via an open-source mentality
- Unlocking collaboration, learning, mentoring and cross-pollination of skills, ideas, and innovation from previously inaccessible knowledge and tools
- Increasing resources for businesses to prototype and test, speeding up innovation
- Allowing for localised design and customisation to foster the pluriverse

It's important to note that for distributive design to be fully sustainable, the materials needed for making the equipment itself, must come from sustainable materials or circular loops. And the designs themselves must embody circular design, with makers considering maintenance, repair, and disassembly for refurbishing and recycling.

At the forefront of the distributed design movement is the Distributed Design Platform in Europe (DDP), co-funded by the Creative Europe program of the European Union.

Established in 2017 by a culturally and creatively diverse group of Fab Labs, cultural organisations, universities, and maker spaces, the DDP supports emerging practitioners of distributed design with collaborative, sustainable, open-source, and eco-systemic values to produce 'good design, smart manufacturing, and quality'. They also produce free design books and toolkits, and they are looking at how to scale their platform to address climate, social, and sustainability issues.

With increasing interest and investment, distributed design has the potential to grow sustainable and regenerative local economies by generating more income for existing businesses, creating new local business and job opportunities, and reducing waste.

# Sustainable digital design

Every web page visited, image loaded, video streamed, online purchase made, information submitted, and backend system call generated by online activity requires energy to power and process the data flowing between devices, network systems, servers and back again, many times over—all of which emit $CO_2$.

Information Communication Technology (computers, phones, and TVs) are now responsible for far more greenhouse gas emissions than the aviation industry[80].

The best place to start reducing the emissions from digital products and channels is to check options for green web hosting.

## Green hosting

Simply altering your choice of hosting can make a significant impact:

- Check who is hosting your site and where they are located using 'hosting checker' online tools
- Browse green hosting options, powered by renewables, via *thegreenwebfoundation.org* (although the cost of green hosting might be a barrier for some)
- Change to a provider close to the target audience, or use a content delivery network (CDN) for sites with a world-wide audience

The next step is to do a sustainability audit on your digital product or channels to determine the best sustainable strategies to improve.

## Audit digital sustainability

Assess for opportunities:

- Ecograder provides a score out of 100 for websites and provides advice on how to improve sustainability via performance, find-ability, and UX— *ecograder.com*
- The Website Carbon Calculator estimates the carbon footprint of web pages and provides tips to improve—*websitecarbon.com*
- To aid continuous improvements, 'Ecoping' is an analytics tool to track website carbon emissions—*ecoping.earth*
- Pagespeed.web provide reports on the performance of a page and provides suggestions on how it may be improved— *pagespeed.web.dev*

There is a lot to improving the sustainability of digital experiences, and much of it is to do with minimising code use and system structures. But there are also numerous strategies which the designers can implement to reduce energy used by systems and users.

- Review the strategies below to select ones that will best improve the issues exposed by the audit
- Prioritise strategies by using a 2x2 grid of feasibility versus impact
- Amplify sustainable design strategies that will impact much of the site by embedding them into the site architecture and style guide

## Optimise content

- Use images sparingly and optimise

  o Use WebP or AVIF format for complex images like photos, and PNG or GIF for simple images like logos
  o The more colour and complexity in an image, the more energy required to load and display it, so choose simpler imagery when possible
  o Use images sparingly, or tell the same story with a few SVG/CSS graphics if possible
  o Optimise the images further using tools like *tinypng.com*

- Use video sparingly and optimise

  o Use WebM instead of MP4 and GIF—WebM has higher compression, so its quality may not be as good as MP4, but it should be acceptable
  o Keep videos as short as possible, and optimise when done
  o Use video sparingly, or tell the same story with a few SVG/CSS graphics if possible

- Icons & logos as SVGs

  o Use SVG
  o Create a sprite sheet of icons—a sheet that combines all icons into one file which is loaded once, instead of every page loading every icon separately

- Use system fonts

  o Use non-system fonts sparingly
  o Use mainly system fonts—these are pre-installed on devices and therefore don't require loading, but they do limit creativity. Below are some of the common default fonts on Windows and Mac (italics indicates Mac alternative fonts), although not all styles are always supported (bold, italic, etc.)[81]:

    ▪ Sans-serif

      ▪ Arial, Helvetica
      ▪ Arial Black, Arial Black, *Gadget*

- Impact, *Charcoal*
- Lucida Sans Unicode, *Lucida Grande*
- Tahoma, *Geneva*
- Trebuchet MS
- Verdana, *Geneva*

  - Serif

    - Georgia
    - Palatino Linotype, Book Antiqua, *Palatino*
    - Times New Roman, Times

  - Cursive

    - Comic Sans MS

  - Monospace

    - Courier New
    - Lucida Console, *Monaco*

**Minimise styling**

- Page styling is created by HTML and CSS files, which increases the page load and reduce speed, so keep them simple
- Use darker colours which reduce screen energy
- Set a page load time limit

**Design for mobile first**

- Designing for mobile first optimises the design for minimum essential content. You can then aim to maintain this minimum for larger screens

**Maximise user journey efficiency**

- Design navigation and journeys to use as little interaction and page visits as possible

**Minimise emails and communications**

- Emails and their content add up the energy used, especially with millions sent around the world every day. Keep yours to a minimum, with minimal content— will appreciate the brevity!
- Audit your email lists regularly and purge non-engagers

**Encourage sustainable behaviour**

- Recommend customers set updates to auto as new versions are often more efficient
- Use behavioural design to encourage other sustainable behaviour (see Foster user stewardship in Design for lifecycle)

- Partner with more sustainable apps—use this Greenpeace Report to check the sustainability of common apps, *clickclean.org/usa/en*

There is a lot more to these strategies, and many other technical and system-based ones, so dive into resources like:

- *thegreenwebfoundation.org*
- *sustainablewebdesign.org*

# Exercise 3—Map the physical aspects of a digital product

Digital products require physical components to support their existence—such as servers for data processing, retail outlets, etc. A photo editing app, for example, require servers and energy sources.

And they drive user need and use of physical devices and services—using an Uber requires servers, energy, smart phones for users and drivers, and vehicles.

The *Phygital Map (*Figure 8 - Exercise 3-Phygital Map*)* is a tool to aid in mapping a high-level view of these physical resources supporting a digital experience. It isn't a detailed or decision-making tool, it's a thought initiator to identify more detailed lines of investigation.

Using the tool might identify resources that you have direct influence over—you can map some of these through their own *Lifecycle Map* to identify opportunities for life-centred improvements.

And for what you don't have direct influence over, you might be inspired to find ways to regenerate them indirectly.

### The Phygital Map layout explained

The map is a time flow of columns from left to right:

- User finds and gets—how the user finds and purchases access to the digital product (sales channels, website, ecommerce, delivery, etc.)
- In use—how the product is used by the user, and any support required during that time
- End of use—how the use of the product ends, what happens to the user's data, etc.
- Maintenance—any other operational activities to keep the digital product active

The horizontal rows capture actions made by the User, Provider, and 3rd Parties:

- User:
  - Digital—the actions of the user in the digital world during their digital experience
  - Physical—the correlating physical components used by the user that enable the digital activity (device, modem, etc.)

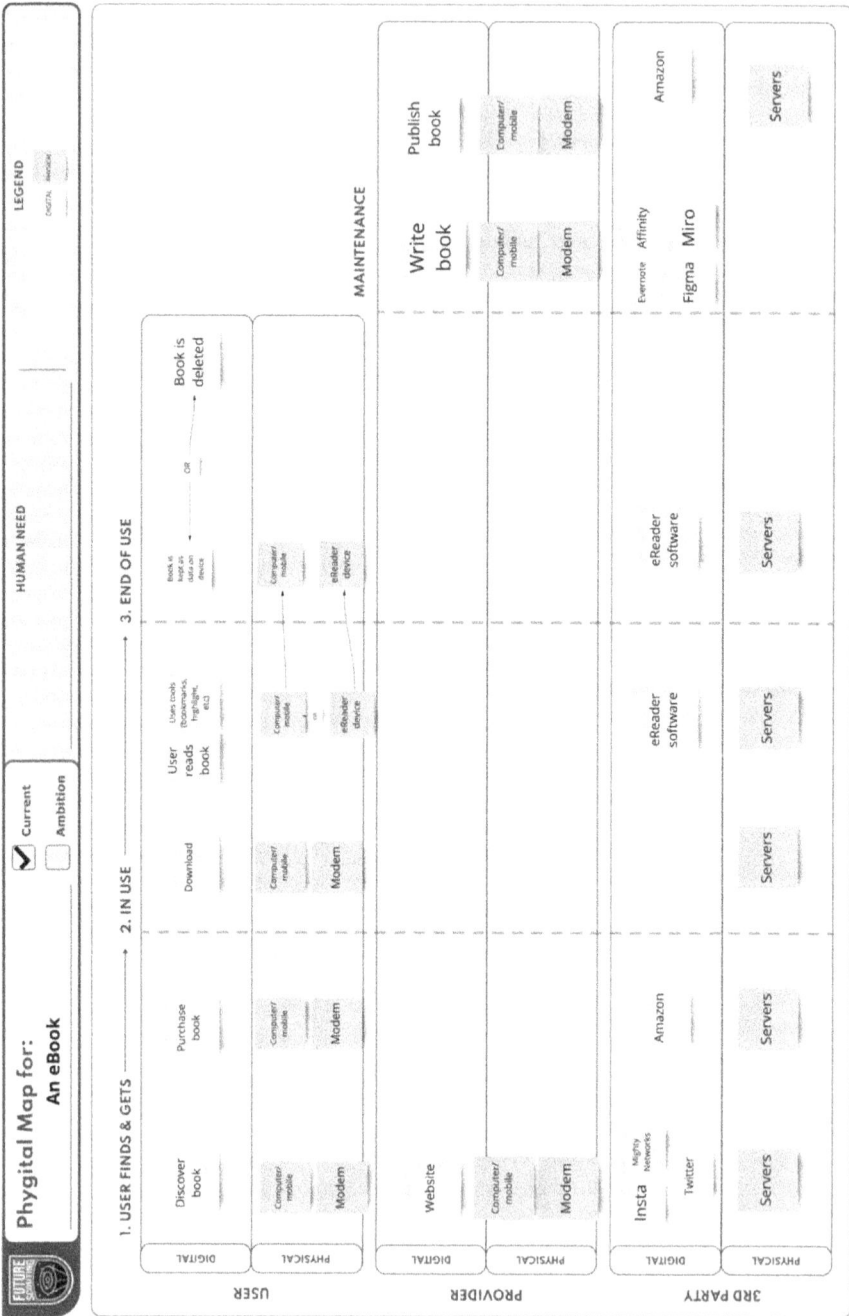

Figure 8 - Exercise 3-Phygital Map

- Provider (Product owner/service provider):
  - o Digital—any digital experience created/enacted by the provider to enable the user's actions (providing website, processing online purchase, etc.)
  - o Physical—the correlating physical components that enable the provider's actions (servers, etc.)

- 3rd Party (any 3rd Party service providers supporting the provider or user's actions):
  - o Digital— any digital experience created/enacted by a 3rd Party to enable the provider's or user's actions (cloud service, PayPal, etc.)
  - o Physical—the correlating physical components that enable the 3rd Party's digital actions (servers, etc.)

This should provide a high-level view of the overall physical component of a digital product's experience. Look at the map through *Lenses*:

- How are *People* affected?
- How will continued use impact the physical resources and people over *Time*?
- Can any components be transitioned to being powered by renewable *Energy*?
- Are there *Relationships* that can be leveraged to explore and foster life-centred collaboration?

Identify any physical and digital components you have direct influence over:

- For physical components, take them through a *Lifecycle Map* to identify *Design for lifecycle* strategies to improve their life-centredness
- For digital components, use the Sustainable and ethical digital strategy in Design for lifecycle

Look at all the physical components again, including the ones you can't influence, and assess for commonalities in resources for focus of regeneration:

- What are the main devices used?
- What are the main materials used in these devices?
- What planetary resources do these take from?

Look at this through the *Regeneration Lens* to explore how your business can give back to these resources

# Behavioural Design

While circular design focuses on what designers, the business, and the supply chain can do, behavioural design focuses on what's in the user's control—where there is often the greatest environmental impact—to inform and enable sustainable and ethical behaviour.

Behavioural design uses a scientific and systematic understanding of human decision making (Behavioural Science) to design products and experiences that influence people's behaviour.

For decades, the commercial world has employed psychological, behavioural, emotional, and social interventions to influence consumers' prioritisation of self over others and the environment. This has also supported the dominant destructive and marginalising power structures.

A 2020 report, however, *How sustainability is changing consumer preferences*, by French consulting and digital transformation company Capgemini, reported 79% of consumers are changing their purchase preferences based on a business's environmental friendliness, social responsibility, and economic inclusiveness[82].

But even with the increase in awareness and sustainable behaviour options, adhering to sustainable behaviour is necessary for it to succeed. Providing information and motivation for users to engage and adhere to sustainable behaviour and resource stewardship is crucial to closing circular loops[83].

## Examples

**Informing:**

- A new online platform in Stockholm offers residents data on their suburb's air temperature, humidity, pressure, wind, rainfall, and solar radiance across the suburb to create a visual reinforcement of how their suburb's 'greening' efforts are influencing local weather conditions, like cooling air during heatwaves[84]
- HEALabel is a free online tool, launched in June 2021, which quickly estimates a product's social and environmental impact and can be filtered by food, material, brand or lists[85]—healabel.com

**Enabling action:**

- An *Uber Eats* sustainable eating enhancement concept by *Good Girls Gang* suggested simple design tweaks such as promoting plant-based meals by running thematic promotions like 'Meatless Mondays' and allowing planet-friendly food to be saved as favourites[86]

- The online grocery app GreenChoice uses data collated from NGOs and third-party certifiers to apply food ratings to 350,000 consumer products found at big-name brands like Walmart and Whole Foods87—*greenchoicenow.com*

There are different behaviour journeys to encourage sustainable and ethical behaviour.

In *Pathways of sustainable behaviour*, Strömberg, Selvefors, and Renström outlined five types of sustainable user journeys[88]:

- User's choice of what they purchase—choosing a product that consumes fewer resources during its use
- Changing their ways of use:

  - o Using the product in a way that consumes fewer resources (energy, data, network usage, etc.) or produces fewer pollutants
  - o Adapting user behaviour according to situations
  - o Curtailing user behaviour that uses too many resources

- User maintenance and repair—encourage and enable both by user
- Mediated use—a second product/service/device is employed to enable the behaviour change in the primary product
- Regulated artefact—a second product, service, or device is employed to regulate the user's behaviour of the primary product

## Ethics

When working with psychology to influence others, however, there is a fine line between positive influence and coercion. Before employing behavioural design techniques, designers must understand ethical design.

Behavioural design is not about deceiving or forcing people to change behaviours. Rather, it is about utilising cognitive biases to create less friction and motivate desired behaviour, while respecting a user's freedom of choice, autonomy, and dignity.

Ethical interaction design means not using dark patterns that leverage base desires to trick users into doing things they don't want to do[89]:

- Trick questions—that appear to ask one thing but ask something else
- Sneak into basket—additional items added to online shopping carts through misleading user interfaces
- Roach motel—baited commitments (subscriptions, etc.) that are then hard to get out from
- Privacy zuckering—being tricked into sharing more private information than intended

- Price comparison prevention—deliberate hindrance to informed decision making
- Misdirection—distracting user's attention from what's really happening
- Hidden costs—unexpected charges
- Bait and switch—being tricked into doing something else unintended and undesirable
- Confirm shaming—the use of shame to influence a user's decision
- Disguised ads—adverts designed to look not like ads, but like curated content or part of the user interface
- Forced continuity—unannounced recharges and renewals
- Friend spam—obtaining access to a user's contacts to spam them whilst pretending to be the user

Berdichevsky and Neuenschwander begun to address these concerns in 1999 in *The Principles of Persuasive Technology Design*, a set of ethical principles to assist in the development of ethical persuasive technologies[90]:

- The intended outcome of any persuasive technology should never be one that would be deemed unethical if the persuasion were undertaken without the technology or if the outcome occurred independently of persuasion.
- The motivations behind the creation of a persuasive technology should never be such that they would be deemed unethical if they lead to more traditional persuasion.
- The creators of a persuasive technology must consider, contend with, and assume responsibility for all reasonably predictable outcomes of its use.
- The creators of a persuasive technology must ensure that it regards the privacy of users with at least as much respect as they regard their own privacy.
- Persuasive technologies relaying personal information about a user to a third party must be closely scrutinized for privacy concerns.
- The creators of a persuasive technology should disclose their motivations, methods, and intended outcomes, except when such disclosure would significantly undermine an otherwise ethical goal.
- Persuasive technologies must not misinform in order to achieve their persuasive end.
- The Golden Rule of Persuasion—The creators of a persuasive technology should never seek to persuade an individual or a group of something they themselves would not consent to be persuaded to do.

Today's thought leaders in digital behavioural design have outlined three simple principles to guide ethical digital design[91]:

- Transparency
- Alignment with social good
- Alignment with a user's desires

## Process

A general behavioural design method follows a design thinking flow:

- Understand
  - o Preform a behavioural diagnosis to understand the current state of behaviour and reasons
  - o Define barriers, blockers, and leverage points that foster desired behaviour and hinder or prevent undesired behaviour

- Design & test
  - o Design interventions
  - o Reduce barriers
  - o Amplify benefits
  - o Form habits
  - o Test and iterate

In 2011, Karin Lidman and Sara Renstörm explored designing for sustainable behaviour in their thesis *How To design For Sustainable behaviour?* and identified categories for design strategies to foster long term acceptability and implementation of sustainable behaviours[92]:

- **Enlighten**—use information, feedback, or means of reflection to influence user's knowledge, values, attitudes,
- **Spur**—motivate a user towards a desired behaviour through expressing benefits unrelated to environmental outcomes
- **Steer**—use physical or cognitive guides to make sustainable behaviours clear and/or default
- **Force**—compel sustainable behaviour through limiting functionality or reducing appeal of unsustainable behaviour
- **Match**—adapt the behavioural design pattern to match the original and expected interaction

Behavioural design empowers designers to identify intervention points and design for motivation uses by understanding how people make sense of information, make decisions, and take action.

## Stewardship

Stahel argued in the 2019, in *The Circular Economy: A User's Guide,* that the shift to circular could be accelerated by 'user-ship'—prosumers and economic actors owning and operating objects in a way that cares for the materials and stocks of objects in their possession[93].

The Australian Government and the European Union, among others, incentivise business-led lifecycle stewardship through policies and legislation. The Stockholm Resilience Centre, authors of Planetary Boundaries, are researching 'biosphere stewardship' to explore ways to amplify and leverage human-nature relationships for social-ecological transformations through 'emphasising notions of care, learning and collaboration[94]'.

But the concept of stewardship—'the active shaping of development trajectories for social-ecological resilience and human well-being[95]' is another life-centred concept with origins in Indigenous worldviews.

In Canada, *Nākatēyihtamowin,* or *Nakaatayihtaamoowinis* is the principle of sustainability as known by the Indigenous Peoples of Canada. It emphasises not just the resources we care for, but also the *relationships* we have with them. They practice stewardship through perspectives such as[96]:

- **All my Relations**—recognising inextricable interconnectedness
- **Seven Generations**—respecting connections through time by always considering the next seven generations
- **Medicine Wheel**—Teachings of balance and equilibrium

In December 2021, Ngāi Tahu, New Zealand's largest southern tribe, began to take New Zealand's government to court, for what they alleged as 'repeated failures over successive governments to protect the country's waterways[97]'. The Ngāi Tahu wanted their *rangatiratanga*—governing authority and self-determination—over the South Island waterways which was returned to the tribe. However, *Rangatiratanga* is not the same as the Western idea of ownership. While it is about governing rights, it is also about being responsible for protection of what is governed.

Contemporary economic alternatives adapt to a stewardship perspective by adopting the Indigenous worldview that all resources are fundamental to our well-being and not to be separated, sorted, sold, and owned, but respected as commons 'to be tended with respect and reciprocity for the benefit of all[98]'.

Stewardship of materials can be extended to users to raise their awareness of the connection between the products and resources they rely on and the peoples, systems, and places that provide them—echoing Indigenous beliefs about the connections between people, and people and land. Designs can foster user stewardship by:

- Utilising transparency of information
- Providing guidance via behavioural design
- Employing pro-sumerism concepts that include users in the production process via direct feedback, feature requests, and/or focus group workshops.

By combining behavioural design with a transparency approach and the Indigenous concept of stewardship, behavioural design becomes more life-centred.

*See the Foster User Stewardship strategy.*

# Biomimicry

Sometime after their arrival in Hawaii, the first Hawaiians learned about the local forests to define those that were sacred and not to be disturbed from those that could be adapted to grow food for humans.

Their 'forest redesign' for agricultural purposes mimicked the native forest's multiple-tiered structure by maintaining canopies dominated by one or two species, planting more diverse and dense sub-canopies, and maintaining dense patches of groundcover of only one specie. This mimicking of nature's systems ensured the agricultural plantings maintained the forest's systemic abilities to maximise nutrients, sunlight, and water, and control flood and erosion, while providing for human needs[99,100,101].

By the 20th century, however, centralised governance and its separated jurisdictions had diluted the importance of maintaining a connected balance between resources. The local forests decreased in size, and residents were forced to increase the importing of food[102,103,104]. Increased urbanisation and growing commercialisation of resource extraction increased the rate of biodiversity loss and environmental damage. The customary practices of Indigenous Hawaiians have persisted, but are slowly being revitalised.

The Indigenous origins of biomimicry—emulating nature to solve design problems—are another history of 'innovative thinking' that often does not reference Indigenous Peoples' ways of living. The famous Italian genius artist of Renaissance times, Leonardo da Vinci, with his studies on birds in the 1480's and how it informed the potential for human flight, is largely acknowledged as a key initiator of biomimic thinking.

Today's emergence of biomimicry in the Euro-Western world marks a transition from the 'era of extracting from nature to learning from its forms, processes and strategies'[105], reconnecting us with natural systems via design.

Put simply, biomimicry is studying and mimicking nature's forms, functions, and systems to inform design with more sustainable solutions[106]. Nature's billions of years of 'research and development' have sustainable and regenerative solutions for many of our problems embedded in its plants, insects, and ecosystems. Designers can study a tree to invent a carbon capture machine, a leaf to improve solar panel design, or the water repellent quality of a lotus to design waterproof fabric.

Biomimicry is applied to the use of energy, chemistry, materials, information management, and town planning, and can be utilised at any solutioning stage of the design process.

Janine Benyus, American biologist, author, co-founder of the Biomimicry Guild and widely recognised as the founder of biomimicry[107], argues that nature has universal trust because all cultures agree that life works[108].

The key goals of biomimicry are:

- Improve human and ecological resilience
- Reduce resource consumption and optimise what is already available
- Regenerate living systems and biodiversity by turning waste into food

### How biomimicry works

The process of biomimicry reduces human design problems to their desired basic functions, which are then correlated with animals, plants, or ecosystems that have solved that same problem (using a resource like *asknature.org*). These biological solutions (forms, functions, and systems) are then reinterpreted into a technological solution for the human problem.

This scientific approach to emulating nature is complimented by an environmental mindset that refers us to nature for a sense of standards, and a philosophical perspective of nature as teacher.

### Examples

- Interrobang from San Jose produced their **CactiShirt** concept for the Biomimicry Institute's Youth Design Challenge 2021-22 which was designed with microscopic folds mimicking desert cacti to increase surface temperature for more heat release, and a waxy surface inspired by the ephorbias plant to repel heat[109]
- The swarm intelligence of birds, bees, and fish inspire collaborative behaviour for robots
- The tiny hooks on burdock plant seeds that adhere them to clothes and fur inspired the hook and loop design of Velcro
- The serrated edge of a whale's flipper inspired serrated-edge wind turbines
- Dragonfly brains inspired the AI for a missile interception system

### Biomimicry examples for digital design

- The 24-hour circadian rhythms of physical, mental, and behavioural changes in most lifeforms are replicated by programs and apps which help users maintain their rhythms by changing the colour and brightness of device screens according to the cycles of the sun
- The way human's brains work as they think, inspired the original input-processing-output model for computers
- The Golden Ratio is a proportion found everywhere in nature, from sunflower seeds and the cochlea of human inner ears to spiral galaxies in space. Architects, renaissance artists, and modern designers have used this

ratio to create the best proportion and balance for their creations. The Ratio inspired the Golden Spiral and Phi Grid design tools used for composing images, web page layouts, and designing logos and fonts

Biomimicry—also referred to as 'nature-inspired', 'bio-inspired design', 'bionics', 'bio-assistance', 'eco-mimicry', 'eco-inspiration', and 'biomimetics'—refers to bio-inspired design for hardware sciences and bioengineering.

Today's biomimicry evolved from bionics. Although bionics is often used today in reference to mimicking human biology to design biomedical solutions, the term was originally coined in 1960, and it then referred to copying and imitating nature.

Janine Benyus popularised the term biomimicry in her 1997 book *Biomimicry: Innovation Inspired by Nature*, calling it 'the conscious emulation of life's genius' to solve design problems and redesign manufacturing processes[110]. Benyus cofounded the The Biomimicry Guild as the world's first bio-inspired consultancy (later becoming Biomimicry 3.8), allowing Benyus to impart nature's genius upon the corporate world, including organisations such as Nike, Levi's, and General Electric.

In 2006, Benyus then cofounded the non-profit Biomimicry Institute which created the awe inspiring *asknature.org*, a comprehensive database of over 1700 biological strategies for biomimicry inspiration, from how to generate warmth, manage waste, manage stress, and repel water to communicating, programming, calculating, teaching, learning, and more.

In 2007, New Zealand author and scholar of environmental studies, Dr Alan Marshall, explored the potential of Eco-mimicry—location-specific biomimicry—in his 2007 paper *The Theory and Practice of Eco-mimicry*[111] and subsequent book, *Wild Design*, to localise inspiration and resource use. Eco-mimicry argues that there can be no sustainability without an understanding and respect for local wisdom[112]. Eco-mimicry is common in disciplines that are directly associated with a specific location, like architecture and urban planning.

This focus on native solution inspiration is also echoed in Biomimicry 3.8's *Genius of Place* process which draws on the interactive *Ecoregions Map* to help designers identify and understand their ecological region for localised biomimicry[113].

## Principles

Along with other collaborators, Biomimicry 3.8 defined the six most common strategies used by Earth's lifeforms and systems. They call these six sustainable and regenerative benchmarks for design 'Life Principles'[114]:

- Evolve to survive
- Adapt to changing conditions
- Be locally attuned and resilient
- Use life-friendly chemistry
- Be resource efficient (materials and energy)
- Integrate development with growth

While Benyus and Biomimicry 3.8 have given biomimicry the greatest exposure through a generous sharing of online learning, nature data tools, speaking, and professional services, other organisations are also championing bio-inspired methods.

Ceebios, a national non-profit network of industrial, academic, and institutional players, is dedicated to the accelerated ecological and societal deployment of biomimicry in France. Also in France, B-Corp-certified design agency Circulab infuses their circular and regenerative work with biomimicry, which they also teach via there online design academy. And German based Biokon is a non-profit association that initiates and supports bionics research (there choice of term for biomimicry) and collaborations between science and industry.

*Bio-inspired, another approach* was an interactive exhibition in France that led visitors on a discovery journey through how living things work and how human design can be inspired by them. The exhibition offered a form of principles in their 'bio-inspired pathways to sustainable solutions[115]':

- **Photosynthesis**—how plants redistribute solar energy to all other lifeforms for the full cycle of life processes from creating new life and growing to decay and decomposition
- **Cooperation** —saving energy and fostering abundance and resilience
- **CHNOP**—prioritising the use of carbon (C), hydrogen (H), nitrogen (N), oxygen (O) and phosphorus (P) as they make up more than 96% of all living matter—they are abundant, have no or minimal toxicity, and can be recycled indefinitely
- **Cycle**—all waste is resource or energy for another life form or system
- **Variability**—diversity of resources and solutions enables adaptability and resilience
- **Local**—most lifeforms and ecosystems utilise only the resources in the location of the system they belong to

- **Sobriety**—the design of lifeforms and natural systems mostly utilises only what they need for all aspects of form and function
- **Biosphere**—the importance of the planet's core systems that enable life to continue and thrive (Planetary Boundaries) and how they influence, and are influenced by, the interdependence of living things
- **Interdependence**—the systemic connection between all lifeforms and systems, including humanity, via biological and chemical reactions

Challengers of biomimicry raise questions about nature's truths being lost through human translation and decontextualization. And, like systemic design, biomimicry requires working with specialists, such as biologists, ecologists, chemists, and physicians which can slow down the design process—Biomimicry 3.8 advise a proper process should include, at a minimum, a designer, biologist, engineer, and businessperson.

However, the basics can be understood and experimented with the help of free online tools like *asknature.org*.

Nature's solutions can be applied at any level, to form, function, or system, and therefor Biomimicry can be applied as a lens at any stage of the design process that requires ideation or solutioning.

*See the Nature Lens.*

# Exercise 4–Look to nature for a design solution

*Tool–Bio-inspired solution tool* (Figure 9 - Exercise 4-Bio-inspired solution). *Download from lifecentred.design*

Choose a simple design problem to solve, like 'My product can't get wet'.

- Reduce the problem to a basic function by considering the function of its solution. Keep it simple, like *building, moving, heating, cooling*, etc.—the basic function should read like it could relate to anything human, machine, or nature. In this example, the function is '*repel* water'.
- **Search** *asknature.org* for biological organisms and systems that perform the same function
- **Explore** how the organism or system achieves the solution with its form, functions, and/or by being part of a larger ecosystem
- **Interpret** the biological solution as design principles
- Using the design principles as a guide, write or sketch out an **engineering solution.**

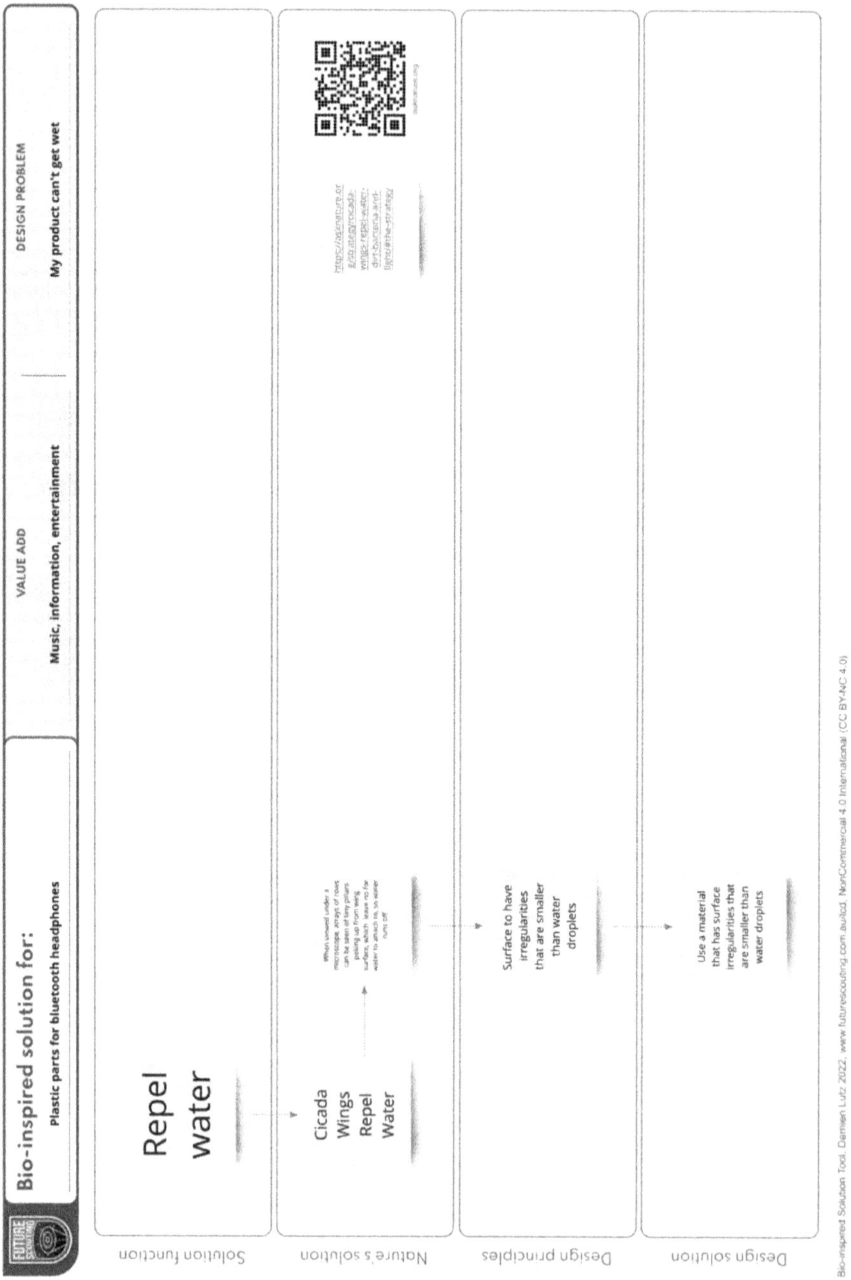

Figure 9 - Exercise 4-Bio-inspired solution

# Foresight

Just as there are many pasts and present ways of being, there are many possible futures as well. Future thinking enables us to anticipate, explore, and test these multiple futures. The overarching goal of futures thinking is to broaden our idea of what's possible, to make mistakes in the virtual realm of future fiction so we can avoid them and inform today's actions to steer us toward our preferred futures.

Only humans can think about and plan for long term futures, but we have neurological blockers that affects the most of this ability.

Foresight and speculative tools enable us to:

- Speculate and explore multiple possible futures
- Explore possible future impacts of today's decisions
- Articulate our ideal futures to start manifesting them
- Identify opportunities for innovation
- Identify and mitigate threats

The multiple possible futures can be understood in terms of types and potentials (Figure 10 - Future types and potentials).

## Key Future Potentials

- **Preferred**—desired states
- **Projected**—the predicted state based on *business as usual*
- **Probable**—likely potential states
- **Plausible**—less likely potential states
- **Possible**—unlikely potential states
- **Preposterous**—not impossible, but highly unlikely

Future types and potentials

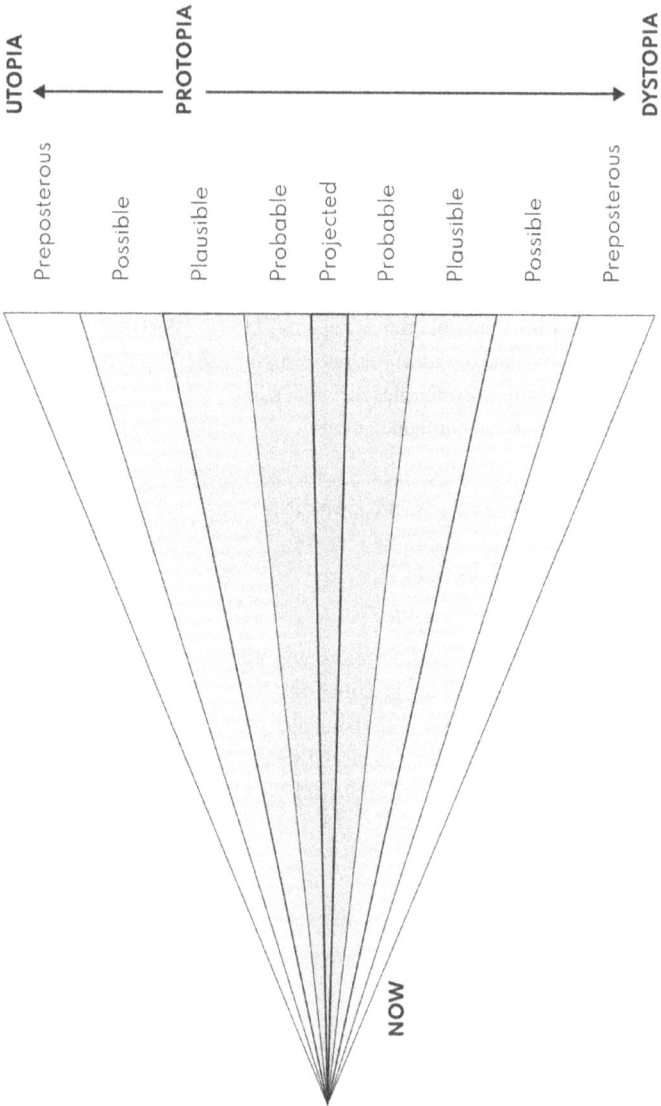

Figure 10 - Future types and potentials

## Key Future Types

- **Dystopia**—science fiction's favourite, dystopias represent the world in worsening states politically, environmentally, socially, technologically, and/or economically, creating unsafe, unjust, unhealthy, and less free futures
- **Utopia**—the opposite of dystopia, where people and planet are free, safe, and thriving. But the problem with the Utopia concept is that everyone has a different idea of what a perfect future should look like
- **Protopia**—a term coined by American thinker and Futurist Kevin Kelly, protopia is the more realistic future type to aim for, as it isn't trying to achieve some impossible perfection, but one that is better than now and always improving

While we should aim for protopia, we should still explore all future types when thinking about futures to be clear on what to aim for and what to avoid.

## Key frameworks

Forecasting predicts one linear future by analysing past and present data to understand trends, from which an informed prediction is made of the future. Forecasting is used by many organisations for strategic and planning purposes, usually for a 3-to-18-month view into the future.

Foresight speculates a range of plausible futures to develop multiple scenarios to challenge and broaden our ideas of what's possible. Foresight was traditionally used in critical and provocative ways but has since become more adopted by businesses, particularly in these unpredictable times. Strategic foresight, for example, is used by organisations to explore how their customers, partners, and business structure might be different in 3, 5, 10 or 20 years in probable, possible, and preferred futures.

The origins of futures studies can be found in war, from governments navigating post-war uncertainties in the 1950s, to corporations adapting techniques in the 1970s as they wrestled with the volatile markets of the Cold War[116]. Today's new war between a sustainable and just existence for all lifeforms and a dominating human culture addicted to exploitation and convenience again raises the awareness and value of futures thinking.

Both forecasting and foresight involve complex skills. Experimenting with foresight, however, can be done with just strong curiosity and open-mindedness, and without complex qualitative analysis.

## Shapers of the future

Common components used to understand the future:

- **Weak signals**—Weak signals are signs of emerging phenomenon with the potential for causing change. Signals can be a 'thing' (technology, service, or product) or a pattern shift (new ideology, change in behaviours or attitudes, etc.); can be found in many forms and mediums, such as news and social media posts, comments and polls; journals/research; data (birth rates, etc.); human, environmental, animal behaviour changes; emerging technologies; behavioural use of technologies; movements, ideologies, attitudes; consumer trends; language trends; innovative services, goods, materials, etc.
- **Trends**—Trends are patterns of emerging signals with enough frequency to suggest they're more embedded than a fad with the potential to disrupt, but nascent enough to be diffused by other trends.
- **Drivers**—Drivers are the large forces steering the world, such as economies, governments, and demographics. Drivers are slower to emerge than trends and hard to change.
- **Wild cards**—Wild cards are rare and low probability events with a high and broad impact range (geographical, industries, classes, etc.). They are imaginable and knowable but hard-to-predict, and they can generate a turning point of a trend or a system. They can be positive or negative, but are not reversible (pandemics and tsunamis, for example, and mass terrorism attacks causing 10,000 or more deaths).
- **Past influences**—Examining the past helps us understand the relationship between past and future by spotting patterns and drivers.

## Scenarios

Speculating multiple future scenarios is useful for exploring how to address uncertain and complex challenges and for testing the potential impacts of today's actions, decisions, behaviours, and attitudes.

Future scenarios are not predictions, but plausible descriptive prototypes of what could happen, and are imagined through speculation from the researched evidence of signals.

Scenarios can be created by applying various tools to ideate how signals/trends/drivers might evolve:

- **Horizon scan** for weak signals of emerging phenomenon with the potential for causing change
- **Make sense** of the signals
- **Look to the past** to identify drivers

- Use the drivers and signals as a guide to **prototype multiple future scenarios**

Mapping multiple scenarios, from ideal to worst case, helps understand the breadth of possibility and informs strategy. One scenario tool is the *4 Arcs*, which is used to brainstorm how futures might evolve on four different trajectories:

- **Growth**—things continue to develop and grow as they do in present day
- **Collapse**—things fail, and life drastically changes
- **Discipline**—things plateau
- **Transform**—an unexpected change occurs (innovation, wild card, etc.) and changes life in unforeseeable ways

## Scope Wheel

Assess the signals you collect regarding their significance to your forecasting subject by using the Scope Wheel (Figure 11 - Scope Wheel).

- In the centre of the *Scope Wheel*, write the subject of your future exploration
- Place your researched signals in their relative categories, using the circular areas of 'primary, secondary, tertiary' to position them in accordance with their potential for impact to the subject

Use this as a guide to know which signals to explore in further detail.

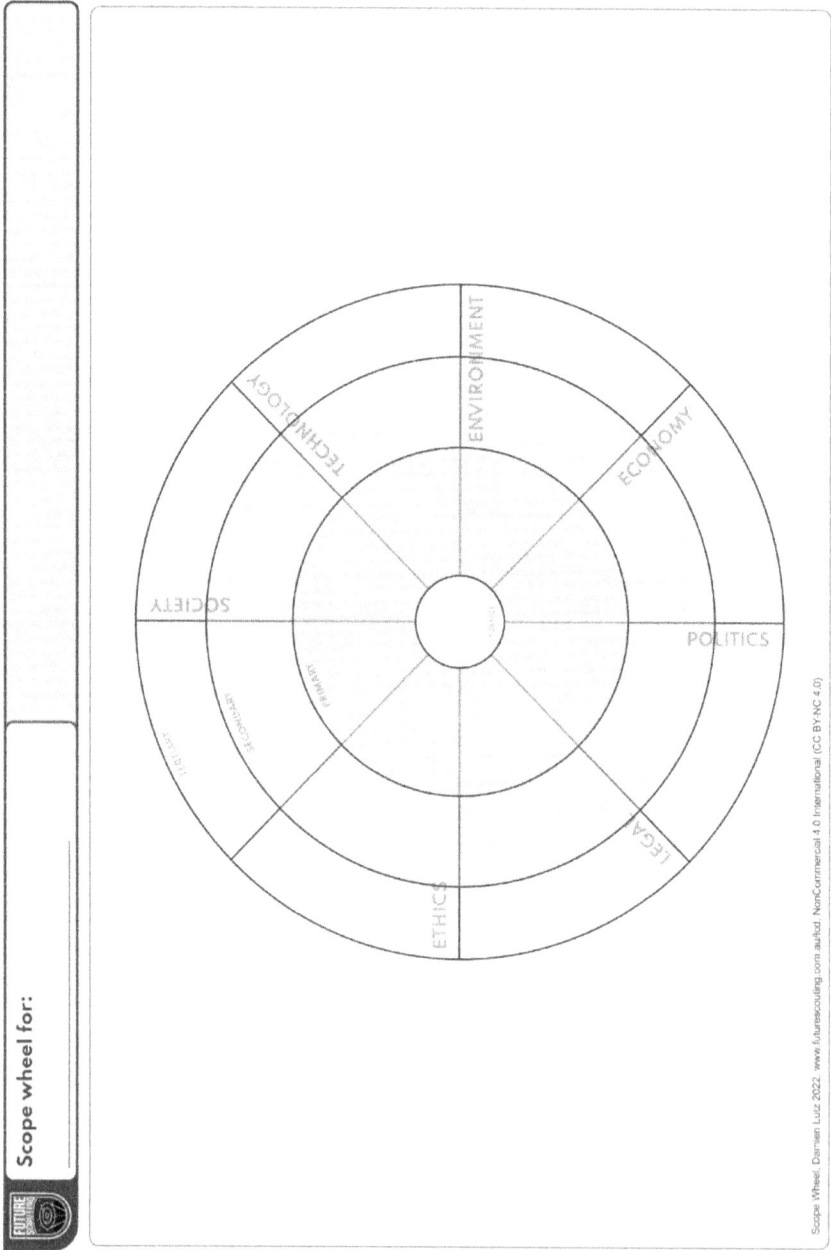

Figure 11 - Scope Wheel

## Back-glance

Understanding the past in its multiplicity broadens our understanding of people and the impact of past decisions and provides essential insight for making better decisions for the future.

Back-glancing identifies key events in the recent past that have a greater probability of affecting the foresight subject.

- Research moments of change (historic events, technology advances, new laws, wars, scientific discoveries, business innovations, etc.). Look back at least twice as far as you are looking forward, but you can go back as far as you feel best
- Arrange each significant change in chronological order
- Review to see what you can infer:

  o What made these changes happen?
  o Any patterns in change?
  o How often between big changes—are they slowing down, or becoming more rapid, and can you infer a rate?
  o What were direct and secondary effects?
  o How have values and needs changed?
  o Do they still impact today?

- Identify significant forces or patterns that will continue to affect the future and summarise into key insights

## Futures Wheel

The *Futures Wheel* (Figure 12 - Futures Wheel) is a brainstorming tool to speculate how the future impacts of design or business decisions (new features, new materials, new business model, etc.) might combine to generate other unexpected impacts.

The Wheel can be used in a variety of ways, but this is a standard use:

- Write your idea/decision/change in the centre of the wheel
- Imagine this change occurs and brainstorm and/or research possible direct results/consequences—positive, negative, or neutral—and write these in the first ring around the centre
- Identify secondary results generated by the direct consequences
- Use the connecting lines to capture what indirect results the combined effects of two direct results might generate (feel free to add more lines to connect more separated ideas/results)
- Extend into new rings of combined and indirect results as desired
- Analyse and sort these ideas into themed insights

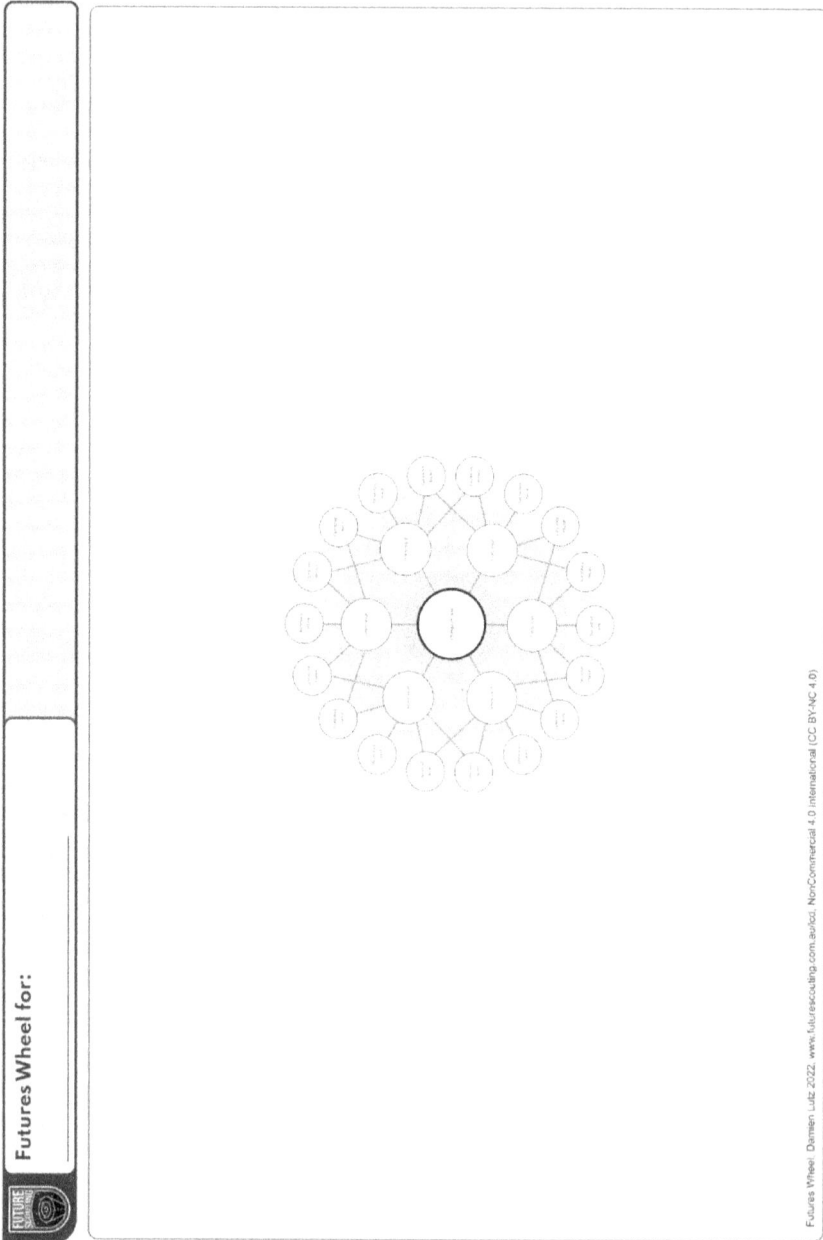

Figure 12 - Futures Wheel

## Future Headlines

*Future Headlines* are fictitious headlines that summarise future scenarios.

Use the Headlines as prompts to innovate product features and/or strategy, to test ideas/changes, and to convey the impact of future scenarios to stakeholders (Figure 13 - Future Headlines)

To create Future Headlines:

- Define the future timeframe and choose a *Future Arc*—Grow, Decline, Transform, or Business as usual
- Research potential future INFLUENCES/RISKS to your product and sort them into groups using STEEP
- Ideate IMPACTS of risks on each group (use the *Futures Wheel*)
- Cluster into themes and summarise as INSIGHTS
- Choose the most interesting/relative insights from each STEEP category and rewrite them as provocative headlines
- Repeat for the other three *Future Arcs*

You can use the *Future Headlines* to explore how your product, customers, business, and design decisions might impact and be impacted in the future.

Used often to assess the impact of ideas, foresight can also form part of an immune system for all peoples, animals, and planet.

*See the Time Lens.*

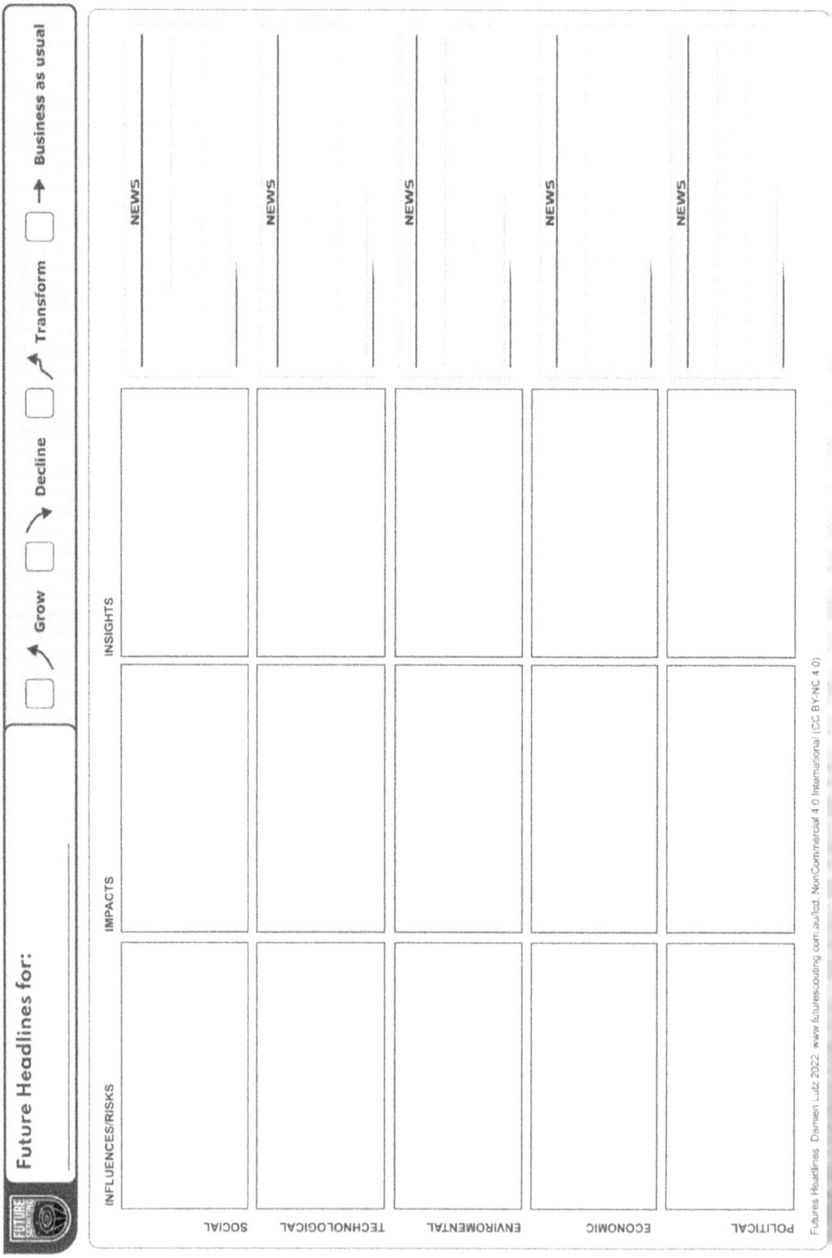


Figure 13 - Future Headlines

# Human-centred product design

Human-centred design is a design methodology that puts human needs and experiences at the centre of all design decisions by engaging those users throughout the process.

By employing empathetic and analogous techniques (simulating the experiences from the user's perspective), human-centred design understands user's conscious wants and unconscious needs and identifies the root causes of problems.

Empathy, 'the ability to understand and share the feelings of another', is the key value that the human-centred design brings to business and design thinking. With empathy as its core value, human-centred design connects users emotionally with technology by delivering to needs, minimising friction, and reducing cognitive load.

Human-centred design has been in popular practice since the late 1990s, and is most often used in conjunction with design thinking, ensuring experiences are optimised for the target human user by following these principles[117]:

- Be human-centred, focus on people, business, and customers, and include in the design process
- Determine and solve the right problem
- Think about the problem as a system and the way the components work together
- Prototype and test

After implementing various techniques, such as contextual observations (visiting people in their everyday environments to observe how they do an activity) and user interviews, designers create prototypes to test them with their users, iterate from the learnings, and then launch their solution into the world.

When collecting insight, two types of data can be collected:

- **Quantitative**—large samples of numbers and statistics about *what* people do
- **Qualitative**—smaller samples of words, meanings, and reasons about *why* people do what they do. There are two types of qualitative insights:

  o Attitudinal—what users tell us about their beliefs, perceptions, and expectations
  o Behavioural—what users actually do, even if it contradicts what they say

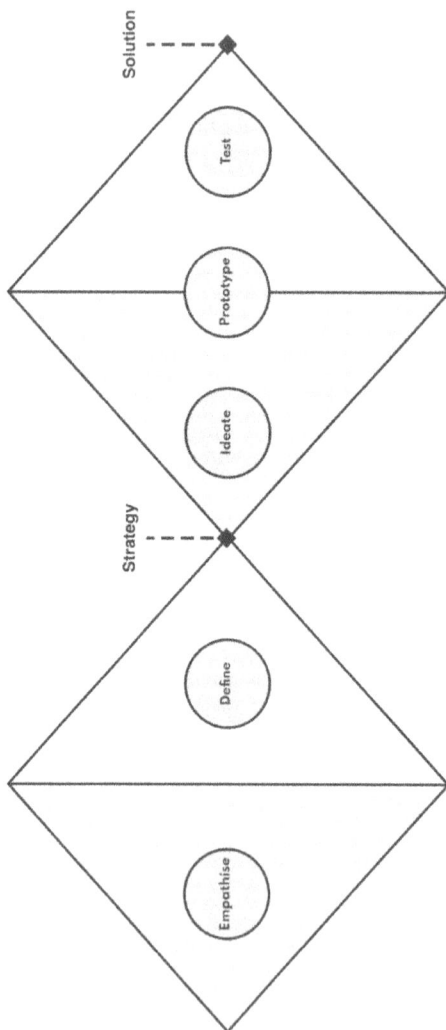

Double Diamond

Design Council

Solution

Test

Prototype

Ideate

Strategy

Define

Empathise

Figure 14 - Double Diamond

## The Double Diamond

Human-centred design is applied across four phases of design thinking innovation, best represented by The Design Council's Double Diamond (Figure 14 - Double Diamond).

Together, the two diamonds represent five steps of the design process.

### Defining the problem and strategy

- Empathise—go wide to gather and understand as much relevant information as possible about the problem and people:

  o Desktop research—web articles, news, academic journals, etc.
  o Collect and review analytics/data
  o Interviews—asking stakeholders and users open-ended questions to probe and empathise
  o Using the *5 Whys*—determining the root cause of a problem simply by asking why to every consecutive answer until the core issue is exposed:

    - **What** is the problem?
    - **Where** does this problem occur?
    - **When** does the problem occur?
    - **Who** is affected by the problem?
    - **Why** does the problem happen?

  o Contextual inquires—observing people in the natural environment
  o Journey Mapping—mapping the user's visual, mental, and emotional experience as they 'journey' through the experience of what is being redesigned

- Defining—Making sense of the discovery to identify opportunities for design:

  o Journey Map—a refined visual representation of users' physical, mental, and emotional experience
  o Personas—fictional persons representing target user groups and their needs and pains to consider through the design process
  o Problem statements—a brief statement defining the background, impacts, and who is impacted
  o *How Might We?* statements—reframe problem statements as broad questions about how to innovate for the specific problem
  o Design strategy—a summary of all findings synthesised into design principles and recommendations

**Exploring solutions and delivering**

- Ideation—Go wide again by ideating many potential design solutions for the user needs and business problems as defined in phase 1:

    o Ideation activities, for example:

        ▪ Crazy 8s—a fast-sketching exercise challenge in which participants sketch eight ideas in eight minutes
        ▪ Card-sorting—giving users words or concepts to organise to guide categorisation, like defining site architecture

- Prototyping—Rapidly prototyping and testing the ideations:

    o Low-fi prototypes—basic representations of the interactive experience in 3D or on paper
    o Hi-fi prototypes—accurate and styled representations of the physical product or interactive experience

- Testing—Rigorously testing and iterating the best prototypes from the previous step to refine the solution:

    o User-testing—giving users multiple tasks to complete using the prototypes and observing for success and pain points to improve

**Each diamond phase is halved into two phases:**

- Diverging—going wide to collect as much relevant information as possible
- Converging—deciding, prioritising, and delivering the solution

The process is designed to be as non-linear as needed, determined by insight gathered from testing with users throughout the design process. For example, testing an initial solution may lead the designer back to ideation, or even back to discovery to re-clarify the problem.

A post-launch step would be monitoring—using data analytics, surveys, and user-testing the live experience and measuring according to success metrics to inspire continuous improvements.

An excellent and well-known example of human-centred design is the MRI experience redesign by industrial designer Doug Dietz[118].

After leading the design and development of high-tech medical imaging systems (MRI) for GE Healthcare, Dietz visited a hospital to view the machine in use.

To his surprise, he discovered the large and noisy machine generated great fear in patients who routinely needed to be sedated.

Witnessing firsthand the anxiety and fear his machine caused, Dietz was inspired to make the MRI experience less frightening for young children.

After learning about human-centred design, Dietz talked to young patients and hospital staff before redesigning the whole scanning experience.

Running a pilot, he applied colourful decals to the MRI machine and gave hospital staff a 'script' to follow, transforming the terrifying and clinical MRI experience into a kid's adventure story with the child as the Hero.

The experience was expanded into the 'Adventure Series' consisting of nine different adventures, from pirate-ships to space travel.

This human-centred redesign resulted in less patients needing sedation, increased process efficiencies, and a significant rise in patient satisfaction—a win for customers and business, and a kinder experience for people when they are most vulnerable.

# Exercise 5—Explore your project's life-centred purpose

*Tool—Life-centred purpose (*Figure 15 - Exercise 5-Life-centred Purpose*)*
*Download from lifecentred.design*

Explore aligning the purpose of your product, business, or personal projects with global goals (such as the United Nations Sustainable Development Goals).

## Page 1

- Complete the header banner:

    o Write the subject (business, product, or project name) at the top
    o Write the value the subject adds to its customers lives, or the value your personal work provides

- In the 'PEOPLES/NON-HUMANS/ECOSYSTEMS' panel, note who and what is impacted by your product/project's lifecycle (refer to your Lifecycle Map from Exercise 1)
- Rewrite these problems as *How Might We?* statements to foster innovate thinking
- Brainstorm different ways your product/project could alter its lifecycle or give back to these peoples, animals, and/or planetary resources.

    o Nurture:

        ▪ All life-centred stakeholders (people, animals, and planet) who might be impacted by your product's lifecycle
        ▪ At all levels—world, organisations/groups, and individuals

    o Think in terms of:

        ▪ Product and service innovations
        ▪ Business model innovations
        ▪ Sustainable and inclusive opportunities within your organisation

    o Use the QR codes at the top right of the page for ideas
    o Think big and small, you'll prioritise later

- Review the global goals at the bottom of the page and tick all that your product/project could contribute to, keeping in mind your product/project's value-add and the ideas you just brainstormed (scan the QR code for more detail on the goals). Try to connect with at least one

goal from each of the three groups (social, environmental, and governance), but just do what feels right and makes sense
- Brainstorm more ideas about how your product/project could contribute to these goals and the ideas you've already generated

**Page 2**

- Sort all your ideas and start merging and refining them. Explore them in a bit more detail and refine them into definite actions.
- Use the FEASABILITY grid on the right of the page to identify the best goals to explore further (the ones that land in the shaded quarter have the most positive impact and are ones you have the highest ability to accomplish). You might also narrow your goals down to a few
- Brainstorm ways to maintain accountability (metrics, deadlines, etc.)

**Page 3**

- Following the prompts, write a life-centred purpose statement to include these goals and metrics. Note your global goals at the bottom for future reference. Remember the key considerations for a life-centred purpose:

  o The life-centred purpose is the higher purpose that the product or service is a part of
  o A life-centred purpose should be succinct in what to do and what not to do, be grounded in possibility but inspiring in its potential. It should guide everything from strategy and operations to culture[119], with diversity, inclusion, and fair pay as defaults, and utilising regenerative concepts to elevate and thrive individual's potentials[120]
  o Finally, a life-centred purpose should be held accountable by establishing measures such as deadlines and timelines to develop credibility and safeguard real progress

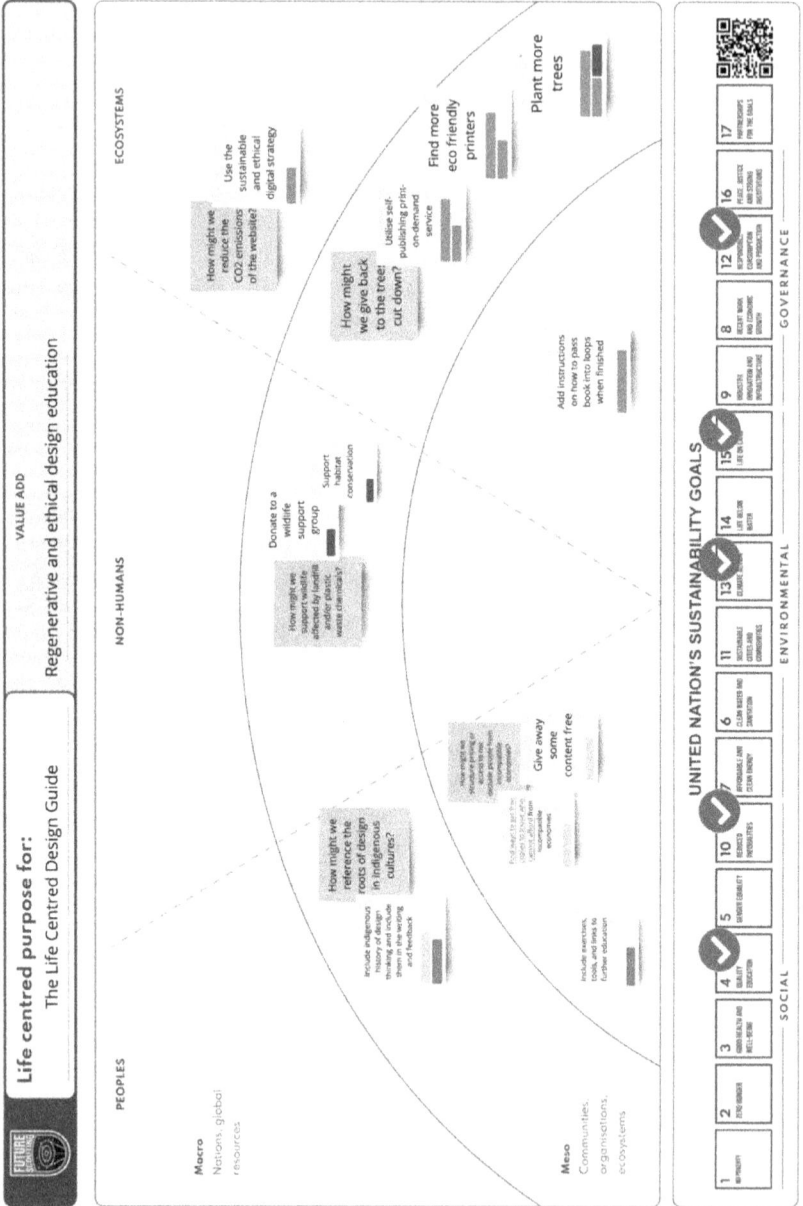

Figure 15 - Exercise 5-Life-centred Purpose

## Alternate global goals

In the spirit of including all voices and perspectives, the Life-centred purpose tool includes a version with the values defined in *Pluriverse: A post-development dictionary* as an alternative set of global goals.

## My example

I used this tool to establish a more life-centred purpose for this guidebook that extended its impact beyond just sharing information about life centred design. You can view this example at lifecentred.design.

# 2.4
# Design
# pillars

This is where the different approaches and various design practices come together.

When the approaches and practices are synthesised, three *key Design Pillars* emerge as life-centred values to uphold to fully consider and protect the *Interdependent Stakeholders*.

These Pillars also answer the three *How might we?* statements about the three problems with modern design discussed in Chapter 1:

## The Life-centred Design Pillars

*How might we design for thriving rather than for continuous growth?*

### Pillar 1—Sustainable and regenerative

- Ensuring the social, economic, and resource sustainability of the product lifecycle by extending material value and reducing waste
- Zooming out to share prosperity with the ecosystem of the product/service lifecycle and to regenerate the sources of energy it takes from, nurturing resilience for all

*How might we design with all the spectrum of humanity, recognising alternate pasts and different ways of living beyond the Euro-Western-centric values embedded in modern design?*

### Pillar 2—Inclusive and pluriversal

- Broadening consideration of human experiences beyond what has been stereotyped as normal within the Euro-Western world
- Zooming out to know and champion diverse ways of existing beyond the dominant Euro-Western-centric capitalistic, colonial, white supremacist, patriarchal, hierarchical, heterosexual, and cis culture

*How might we design with more than humans in mind, considering the impact of our creations upon the land and others across the entire product lifecycle, for now and in the future?*

- **Pillar 3—Responsible and Aligned**
- Align workplace values with life-centred purpose—from individual employee to business ownership—and embrace transparency and accountability to drive innovation
- Convert product to service, distribute to localise, and utilise data to foster responsible stewardship of resources
- Zoom out to align business goals with global goals and ally with the supply chain to nurture circular, regenerative, and just experiences

# Pillar 1—Sustainable and Regenerative

*How might we design for thriving rather than for continuous growth?*

As established earlier, a core practice of life-centred design is circular design, designing for The Circular Economy, which seeks to redesign how we produce goods and services so that they fulfil their intended purpose and meet human needs in more sustainable and regenerative ways.

Sustainability is about perpetuating today's business models in a way that ensures enough resources remain for future generations to get by. But these models contribute massively to today's wicked problems, and 'getting by' is not *thriving*. Any future sustainable design must also heal the damage that has been done, socially and environmentally, and regenerate what it takes from.

Regenerative thinking focuses on giving back much more than is taken[121], generating positive impacts for the environmental and human systems that design and manufacturing activities support or draw from, so that humans co-evolve harmoniously with nature.

Traditionally, Indigenous Australians only took from areas that were bountiful. Decisions to move location were based on their knowledge of plants, animals, and cycles of nature to determine when to hunt, trap, fish, and harvest from when to leave an area to let it regenerate.

By not taking more than needed—leaving honey for the bees to work with, leaving enough seeds to make sure new plants will grow, not hunting young animals or taking all the eggs from a nest—and using any waste as a resource for other natural process—such as burning unused harvested plant parts to enrich the soil—gave areas Indigenous Australians took from time to regenerate season after season[122].

By favouring shared abundance over competitive scarcity, and connecting materials, sources of supply, user experience, and business models with the ecosystems they interact with, businesses can transform products and services that shock, drain, and poison natural systems, into symbiotic components that nourish instead.

This, in turn, fosters resilience for all and generates innovation.

A producer of smart watches, for example, might shift from being just a maker/seller of products to also being part of the local communities supporting its factories, and the ecosystems surrounding them—not just enabling fair treatment for workers in the factories and reducing pollution, but also investing in a thriving local community around the factory and improving local environmental health.

In his 2015, John Fullerton (considered the godfather of Regenerative Economics) explored the idea of a 'self-organizing, naturally self-maintaining, highly adaptive' form of capitalism in his paper, *The Regenerative Capitalism*. He highlighted 8 principles for the regenerative business mindset:

- Robust circulation
- Seeks balance
- In right relationship
- Views wealth holistically
- Innovative, adoptive, responsive
- Empowered participation
- Honours place & community
- Edge affect abundance

Today, many businesses and organisations are exploring sustainable and regenerative solutions:

- **H&M Group** have increased recycled material use to 57%, increased collection of used garments, and begun recycling water from production processes, as part of their strategic vision to become 100% circular
- American company **Dell** plans to be circular by 2030, and are exploring innovative ideas like speeding up the disassembly process by adding a feature to trigger their laptops to 'self-disassemble; making materials more durable; adding AI to manage efficiency; reusing components from old models in new models; and delivering skills to unrepresented groups to access the benefits of technology
- The vertical forest apartments towers of **Bosco Verticale** in Italy houses thousands of shrubs and trees that convert carbon to help with local air quality
- UK based **Econic Technologies** take carbon dioxide out of the environment and convert it into a plastic that can be used for sneakers, pillows, coatings and more, reducing the need for new raw materials

The UK based regenerative action initiative, Positive, define regenerative businesses as 'guided by a transformative evolutionary purpose serving the wellbeing of both people and planet, doing no harm, and making a net positive impact[123]'. Positive connects 'change makers, practitioners, and citizens to grow the field of Regenerative Entrepreneurship' by applying four values:

- Right relationship with nature
- Elevating human potential
- Empowering communities and places
- Value for all

Daniel Christian Wahl, author of *Designing Regenerative Cultures* suggests, however, truly regenerative businesses aren't possible within the confines of our current degenerative economic system, and that they may need to start as a global co-operative of smaller bioregional and local scale initiatives[124]. Even Yvon Chouinard, founder of Patagonia and initiator of its strong ESG commitments, has reflected on his company's billion-dollar scale being too big to ever truly be regenerative. Wahl further suggests that businesses in their current form may need a 'death' to be transformed into regenerative solutions. While some forms of regeneration can take years or decades, this is even more of an incentive to start immediately.

Regardless of how this transformation takes place, leaving things better than when we first encountered them is what will make the world more resilient and adaptable to the unpredictable changes that scientists say await us in the coming years.

A starting point is to take a systems view of a supply chain, identify common resources, and develop ways to nurture and nourish them.

## Sustainable & Regenerative aims and principles

### Key aims to fulfil this pillar:

- Ensure the social, economic, and resource sustainability of the product lifecycle by extending material value and reducing waste
- Zoom out to share prosperity with the ecosystem of the lifecycle and to regenerate the sources of energy it takes from, nurturing resilience for all

### Principles to follow to fulfil these aims:

- **Design out waste and pollution**

  o Circular design reduces waste and pollution by substituting fossil and critical materials with reusable materials recovered from products already made, which also reduces waste and pollution created from mining virgin materials. But circular design is about more than recycling—in fact, recycling is the last resort

- **Keep products and materials in use**

  o Keeping products in loops of reuse, repair, renew, and recycle as it extends the value of those materials

- **Regenerate environmental and human systems**

  o The circular economy's aim is not just reducing harm. Creating products and services that exist as symbiotic components of other systems also means designing in ways that heal damage and further enriches both environmental and human systems, moving sustainability from 'do less harm' to 'do more good'.

# Pillar 2—Inclusivity and the pluriverse

*How might we design with all the spectrum of humanity, recognising alternate pasts and different ways of living beyond the Euro-Western-centric values embedded in modern design?*

There are at least 64 genders, 46 types of sexuality, over 3800 different cultures, more than 6900 languages, and at least 7 types of disabilities. Over 1 billion people (about 15% of the world's population) have some type of impairment that affects their ability to properly access the web, and every non-disabled person will experience a form of disability at some time in their life, whether from injury, illness, or old age.

While there's a lot to understand about designing for accessibility, it's not about retrofitting designs but designing for all people by default from the beginning.

But as mentioned earlier, the lifecycles of many products created by the Global North impact the workers in mines, farms, and factories in the different countries of a supply chain, and the communities nearby, including peoples in the Global South.

Whether physical or digital, the demand for products in the Global North drives injustice in the Global South. Computers, laptops & mobile phones, for example, rank in the top 5 of The Global Slavery Index's list of consumer goods areas most at risk of slavery in their supply chain, which contributes to over 40 million people trapped in modern slavery worldwide.[125]

To protect the full range of diversity connected to a product's lifecycle, inclusive life-centred product design needs to have an awareness and respect for ways of being that differ from the Euro-Western-centric focus on constant development and financial growth.

This is where pluriversal design can inform inclusivity as it is shaped for life-centred design.

While pluriversal design is practised more at a community and global level, it becomes more relevant to product and service design when we consider the full product lifecycle and the various cultures, social systems, locations, and ecosystems which the lifecycle impacts.

Also, digital and product designers are often creating experiences for mass use by multiple cultures, so any small exclusion or marginalisation can be greatly amplified and perpetuated. This risk is being further amplified by emerging technologies, such as the virtual and augmented realities of the metaverse, an innovation already enabling monopolisation and privatisation of digital space and experience.

By adopting a reflective and 'un-learn to relearn' approach, and interrogating their relationship with cultural knowledge, designers at all levels can contribute to the decolonisation of design[126], and identify where exclusion and social injustices are hidden in the design process, and in themselves.

While not every simple product or digital experience might require rigorous examination for perpetuation of marginalisation, many designers of privilege are often not aware of their privileges to know their impact. By re-educating designers about the power of design and its historical political and social impacts, pluriversal design can inform product design principles and tools with global social innovation mindsets to design for a world growing more digital, diverse, and interconnected.

These interconnections highlight how a thriving humanity depends on a thriving planet for the air we breathe, the soil we grow food in, and the waters we drink from[127]. Recognising the intertwining of global ecosystems with the social foundations of our societies reconnects us with those two relationships which the Indigenous Peoples tell us are the most important for all humans—that is between people, and the relationship between people and land.

Businesses need to play their part, too, taking example from companies like Patagonia, who created their manifesto about becoming an 'anti-racist company' and who share their ethical footprint to remain transparent and accountable[128], and Dell, who embed circularity and social justice in their business model.

By drawing from the grounding reminders of alternate pasts, and incorporating the design innovation of pluriversal thinking, inclusive design becomes truly inclusive by awakening modern product design to inherent biases and marginalisation.

While the influence designers and businesses have is limited, awareness of these impacts can generate ideas for redesigning products and supply chains that not only reduces the negative impacts to human diversity, but also heals damages done.

With physical products and digital experiences reaching more global audiences, and life-centred design expanding consideration to include the impact of the lifecycle on people and the planet, pluriversal design gives designers the tools and mindsets to look beyond today's scope of inclusion.

'Pluri-clusive' thinking acts as a grounding pause to:

- Educate designers about the inherent and hurtful biases in modern product and service design, and in themselves
- Raise awareness to spot dominance and injustice at any level
- Empower designers with an appreciation for alternate ways of knowing the world
- Enable designers to decentre themselves from the process so they can positively shape the human experience by designing for the thriving of diverse and intersectional ways of being

To connect today's common idea of inclusive design with the pluriverse, the guiding principles and methods from both inclusive and pluriversal thinking were merged. The resulting principles for micro-level 'pluri-clusivity' listed below can also be adapted to more generalist problem-solving at the macro-level.

While this was not a fully comprehensive examination, this summarisation of over a decade of experience from multiple disciplines and cultural perspectives should at least be a helpful introductory guide for the purpose of practical experimentation.

## Pluri-clusive aims and principles

**Key aims to fulfil this pillar:**

- Broaden consideration of human experiences beyond what has been stereotyped as normal
- Zoom out to know and champion diverse ways of existing beyond the dominant capitalistic, patriarchal, hierarchical, white supremacist, fully abled, heterosexual, and cis culture

**Principles to follow to fulfil these aims:**

- **Know context**

  o Be mindful of design's influence on communication, culture, and power
  o Be mindful of the dominant narratives (white, colonial, capitalistic, patriarchal, hierarchical, heterosexual, cis, fully abled) to better see and understand alternate values
  o Know the context of other's experiences (users, partners, team, invisible humans—all peoples)
  o Be mindful of one's own values, place in context, and any privileges, and use these to aid others

- **Gather the right people and let them design**

  o Establish and facilitate collaborative and participatory design early and often
  o Recognise past knowledge, wisdom, innovation, and ways of existing as a human
  o Allow the design process to adapt and evolve according to the uniqueness of the people and project
  o Ensure they are included at key decision making and testing times[129]:
    - Set up
    - Diverge
    - Converge
    - Wrap up

- o Use various forms of interactive practices such as spoken, visual, artistic, ritual to foster visceral engagement
- o Poetry, music, sand drawing, sand structure building, writing, storytelling, world-building, rapid prototyping, animations, video, gamification
- o Create visuals of problems, desires, archetypes
- o Use mediums that foster creative thinking—dice, cards, Lego, string, etc.

- **Solution for inclusive, adaptable, sustainable, joyous, and interdependent experiences**

  - o Aim for solutions that are as accessible, easy to use, adaptive, and multi-modal as possible to benefit many
  - o Design for sustainability that is also regenerative and considers animal ways of being
  - o Design for more than basic needs to foster a just, equal, joyous, and thriving solution
  - o Increase autonomy by fostering locally defined, locally resourced, anti-hierarchical, self-governing social structures
  - o Design channels and mediums for healthy interdependence with greater eco and human systems
  - o Foster the celebration of equality, inclusivity, and differing ways of existing in the process and solution
  - o Consider future scenarios and stakeholders
  - o Monitor solutions to stay with the problem

# Pillar 3—Responsible & Aligned

*How might we design with more than humans in mind, considering the impact of our creations upon the land and others across the entire product lifecycle, for now and in the future?*

The third design pillar, Responsible and Aligned, can be mostly applied to the business model.

While design is key, decisions made before design begins are even more important as they define the business model which can limit the life-centredness designers can achieve.

Responsible businesses align their goals with global goals and align with the supply chain to nurture circular, regenerative, and just experiences. They zoom in to align workplace values with their life-centred purpose, and they hold themselves accountable with data and transparency.

Drawing from the great work of Impossible, Vincit, The Design Council, DEAL, The Ellen MacArthur Foundation and IDEO, Circulab, Disruptive Design, and others, the following strategies aid businesses in becoming more life-centred-responsible:

- Align with global goals for a regenerative purpose
- Strengthen with transparency
- Convert product to service
- Partner with value
- Distribute to localise
- Life-centred culture

## Align with global goals for a regenerative purpose

Businesses often aim for the biggest market, but these may not be the best in terms of life-centredness[130].

And while many businesses state a clear purpose and values, defining a life-centred purpose conveys how a business aligns making profits with generating environmental and societal value.

Life-centred businesses transition from being producers of goods with little responsibility of lifecycle impact to actively taking responsibility for as much of the lifecycle as possible. They can be mechanisms to align partners, communities, individuals, and ecosystems with global goals by defining and sharing a clear social and environmental purpose, from the business model and circular design of their product to the inclusivity of their workplace.

Experts in the field highlight the importance of defining a life-centred value proposition by focusing on adding value to the system, not the product alone[131], by aligning with the global goals and setting metrics with respect to the ESG framework (Environmental, Social, and Governance measures).

Three great examples:

- Philips—Along with utilising EcoDesign, Philips partner with suppliers to reduce the environmental impact of the entire lifecycle and to foster fair and dignified working conditions
- Patagonia—Support fair work conditions across the supply chain; use regenerative materials; give 1% of sales to environmental protection, foster reuse of their products via the Worn Wear portal; and embed social justice in the workplace and governance
- Dell—use sustainable materials and other circular strategies; cultivate inclusion in the workplace; embed ethical culture and values in governance; and upskill underrepresented groups

While there is no one consistent methodology yet for measurement, traceability and accountability, or a consensus on just how to properly link business-orientated ESGs with global goals, committing to ESG ratings is already compulsory in some countries, and there are a growing number of third-party rating services that can add credibility to business's ESG stories through sustainability performance monitoring and life-centred 'scorecards'.

As standardisation and regulation advance, both consumers and investors are becoming more aware of using them as metrics to guide spending and investment, and thereby accelerating the transformation towards a regenerative and sustainable one in which life-centred businesses will thrive.

An interesting mapping by the European bank Berenberg, shared in Understanding the SDGs in a sustainable investing, categorised the UN's sustainable development goals into each of the environmental, social, and governance categories[132]. Another interesting grouping is used by Impossible by sorting the goals into People, Planet, and Profits (Figure 16 - SDGs mapped to ESG).

But having a life-centred purpose is not just about ticking boxes.

*The Positive Handbook for Regenerative Business* argue that only businesses including nature in process, and appreciating true wealth in environmental, social, *and* financial, are in the 'right relationship' with nature.[133]

# SDGs mappings to the ESG framework

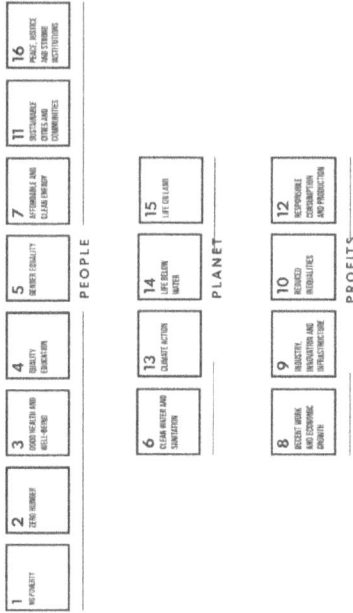

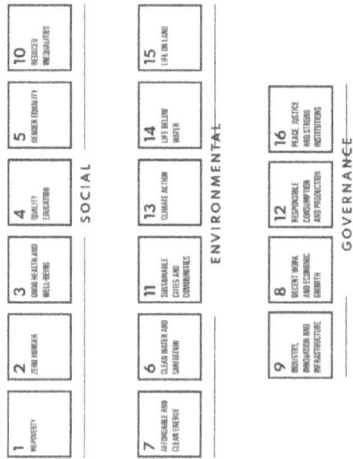

Figure 16 - SDGs mapped to ESG

Below is an example of Philips' life-centred in an ESG framework[134].

- Environmental
  - Maintain carbon neutrality and use 75% renewable energy in our operations by 2025
  - Generate 25% of our revenue from circular products
  - Circular economy—generate 25% of our revenue from circular products
  - Embed circular practices at our sites and put zero waste to landfill by 2020
  - All new product introductions will fulfill our EcoDesign requirements by 2025
  - Work with the suppliers to reduce the environmental footprint of our supply chain
  - Engage with our stakeholders and other companies to drive sustainability efforts

- Social
  - Improve the health and well-being of 2 billion people per year by 2025, including 300 million people in underserved communities
  - Lead with innovative solutions along the health continuum
  - Be the best place to work for our employees, providing opportunities for learning and development, embracing diversity and inclusion, and assuring a safe and healthy work environment
  - Improve the lives of 1,000,000 workers in our supply chain by 2025
  - Actively engage and support the communities in which we operate

- Governance
  - Combine management structure and governance with responsible leadership and independent supervision
  - Being transparent about their plans, activities, results and contributions to society (e.g. tax reporting), and engage with shareholders, customers, business partners, governments and regulators through a variety of platforms
  - Ensuring ethical behaviour through their General Business Principles, with a strong compliance and reporting framework

Each new generation is more aligned with their values and let their values guide the decisions they make, from shopping to whom they work for. Life-centred business will shift the focus from the number of units sold (which increases use of resources) to numbers of consumers subscribed due to valuing how the business aligns with global goals.

**Key considerations for a life-centred purpose:**

- The life-centred purpose is the higher purpose that the product or service is a part of
- A life-centred purpose should be succinct in what to do and what not to do, be grounded in possibility but inspiring in its potential. It should guide everything from strategy and operations to culture[135], with diversity, inclusion, and fair pay as defaults, and utilising regenerative concepts to elevate and thrive individual's potentials[136]
- Finally, a life-centred purpose should be held accountable by establishing measures such as deadlines and timelines to develop credibility and safeguard real progress

## Strengthen with transparency

Consumer expectation for transparency from organisations rises every year[137], and demand for credentials is increasing. But consumers don't expect businesses and their products to be perfect—they just want them to be honest, transparent, and always improving.

Life-centred businesses and designers can look at how they may transform less favourable information into tools of honesty and transparency as a commitment to growth, and as a means of attaining certifications to earn more customer trust.

Transparency is about sharing information that brings value to others, even if it doesn't benefit the business:

- Being open about business flaws with commitments to improve, and sharing commitments to show accountability, fosters honesty and trust from all
- Sharing data with business, designers, customers, and the supply chain via intelligent and connected products, services, and systems allows monitoring for collaborative and continuous improvement
- Transparency can assist in attaining certifications in sustainable and ethical standards.

There is also growing consumer demand for transparency of the source of resources used in products.

The fashion industry, known for its impact of planet and people, is now employing a near real-time, traceability system that provides insights from the start to the finish of the manufacturing process. Traceability is the ability to track products as they move through the lifecycle, including origins of materials and processing history. It also supports sustainability, fair work conditions, and biosecurity. For example, if a shopper can use a QR code to check a product for its sources, they're empowered to consume in alignment with their values and contribute to the development of sustainable and fair supply chains.

However, there are risks to exposing your specifics[138] to keep in mind:

- Allowing others to copy your success (but your USPs and practicing of values can still set you apart)
- Be subjected to analysis out of context
- Sharing and managing response can distract business focus from higher priority work (so keep a balance of time invested in transparency work—start small and keep it manageable)

Fundamentally, being transparent internally and externally with reports, pricing, customer data handling, salaries, open-source code, values, roadmaps, etc. signals to consumers, employees, and partners a responsible and honest organisation[139].

## Convert product to service

A key circular strategy is taking a step back to review how individual 'ownership' of a product can be converted to 'access' via a service model.

Mass production of physical products requires massive amounts of materials and energy. A service-first approach seeks to minimise resource use by reducing individual ownership through converting a product to a service and digitising the experience as much as possible.

Types of service models to consider:

- Renting
- Subscribing
- On-demand access
- Leasing
- Sharing

Transforming a product into a service correctly can significantly reduce these impacts.

This is not to be confused with adding a service focus to your products, keeping your products as must haves with a subscription model attached. This is about removing or reducing the number of products you create and circulate by renting, leasing, sharing, or subscribing the use of these fewer items. However, adding a service focus to existing products can add sustainability if it allows the manufacturer to retain ownership of the product.

When the business retains ownership of its products, it is responsible for maintenance and repair, and this can allow the business to influence more of the lifecycle with life-centred thinking. Designers and product owners can then work closely with manufacturers and suppliers to troubleshoot, adapt, and innovate design to maintain circular integrity.

Exploring how else a human need can be met other than by an individual owning something can also innovate business models and industries, and generate new revenue streams, new roles, and alternate uses for technologies.

Service based models are also beneficial for technologies and industries that are evolving quick, as it allows for faster adaption and less waste of older, un-upgradable products.

Keep in mind, blurred lines of ownership can risk a lack of user awareness of their responsibility[140]. And not all product-as-a-service models remain sustainable, such as the multiple bike-sharing businesses that failed and left behind mountainous piles of unused bikes.

**Service model examples:**

- Philips' transitioned from selling light bulbs to providing lighting-as-a-service, saves customers money by only paying for the light they use and puts the waste of burnt-out bulbs in Philips' hands to properly manage and reclaim valuable materials
- The Netherlands-based 'Bundles' business offers subscriptions on white goods and coffee. They deliver and install machines in a subscriber's home, foster sustainable use and provide sustainable accompanying products, and provide ongoing maintenance (user's existing appliances can be taken away for recycling). Customers simply pay a subscription free which is calculated on usage
- Video, gaming, and music streaming subscription services for streaming digital entertainment replaced the owning digital files or CDs

## Partner with value

Life-centred businesses aim to impact every aspect of their product and service, creating new relationships and opportunities, giving businesses more ability to manage system views, material selection, etc. to steward the resources they use.

Improve the circularity of your product by changing mindsets:

- From 'supply chain' to 'value web'[141]
- From 'competition' to 'collaboration'

Improve the sustainability of the linear supply chain by thinking of it as a webbed ecosystem transferring and sharing value, from material extraction and manufacture to transporters, wholesalers, retailers, and material recovery systems. It is less about profit and cost-cutting and more about creating efficiency and resilience through leveraging shared value. Connecting silos by shared value compounds life-centred impact, generates innovation, builds trust, awareness, creates respect between partners, and embeds human rights more deeply in the value web[142].

Leverage the shared value by collaborating with the entire supply chain and ecosystem, from material extraction and manufacture to transporters, wholesalers, retailers, and material recovery systems. Whether you're in direct contact with each actor in the entire lifecycle of your product, you are in a partnership with them, identify and foster opportunities to collaborate, innovate, and strengthen programs.

You can also partner with organisations doing similar work, have a similar life-centred purpose, or work with similar people.

Going beyond the shared value, local businesses can connect outside their industry with Industrial symbiosis—converting waste, by-products, or surplus resources into usable raw materials for other industrial processes, maximise value of materials, reduction of energy and pollution, innovation of new revenues and processes, and foster long-term cultural change. This can also help transition businesses towards more decentralised and distributed futures. A well-known example is the Kalundborg industrial park in Denmark where nine industrial organizations share and reuse 25 different flows including water, energy, and materials[143]. And, on a smaller scale, Portuguese energy company EDP sell their slag to cement companies to be used in pavements[144].

## Distribute and localise

At the time of writing, the supply chains of many goods in Australia have been slowed or stopped due to the Queensland floods, with potentially more impacts from the rising price of oil from the Russia/Ukraine war. In some areas, the supermarket shelves for eggs, fresh fruit, and vegetables sit empty after only recently recovering from the impacts of COVID19 and decreased availability of workers.

Meanwhile, in the Middle East and North Africa, the rippling effects of the war between the two countries that supply a quarter of the world's wheat has caused prolonged supply-chain disruption and price hikes of bread which make up over half of the calorie intake for the average household in countries like Yemen.

These unpredictable events exposed the fragility of global value web reliance on centralised businesses supplying non-localised sources and how we are only 'one fridge or pantry away from panic buying'.

In conjunction with community-driven changes, like transforming decorative home gardens and community grassed areas into food gardens, and implementing a trade and barter system with neighbours, businesses can localise the design process, resources.

For physical products across multiple regions, production can be digitally distributed to various maker shops and/or fablabs that can produce supply for their local regions on demand. This reduces the materials and energy used in transporting and storing of stock that may never be sold or used, reduces energy and waste from extended supply chains, increases business resilience, and fosters a pluriverse of resilient, autonomous, interdependent, and ecologically minded communities[145].

When used as part of a business model, distributed design spreads design and manufacturing across multiple 'nodes' of digitally connected, citizen-led manufacturing spaces. This moves data instead of products, as an alternative to current centralised mass-production and long, energy-intensive supply chains.

On a deeper level, diversifying through localising reconnects people with place, redefines the relationship between people and the environment, and decolonises the design process.

## Life-centred culture

Life-centred business culture aligns governance, ownership, finance, and workforce diversity with the life-centred purpose to safeguard the purpose and roadmap over time through pressures that might challenge the commitment[146].

Diversity in workforce is going beyond festivals and one-day-a-year recognition ceremonies. It's about fostering a perpetual culture of empathy and awareness of power, privilege, and bias[147].

Discrimination is embedded in all of us to different degrees, and it flows over into our biases at any time—when hiring, setting expectations, reviewing achievements, and considering people for opportunities, etc.

There's an argument that employers shouldn't employ someone just for diversity's sake, especially over someone else who might have better skills, but perhaps we should value diversity of human experience as important as having certain skills and levels.

A diverse workforce not only fosters inclusion, but also brings other benefits[148]:

- Gain access to representatives of diverse users for innovation, decision making, design, and testing
- Attract a wider interest from potential employees
- Attract a wider customer base from the diversity embedded in the business profile and product
- Increase productivity[149]

## Responsible & Aligned aims and principles

**Key aims to fulfil this pillar:**

- Align workplace values with life-centred purpose—from individual employee to business ownership—and embrace transparency and accountability to drive innovation.
- Convert product to service, distribute to localise, and utilise ethical data to foster responsible stewardship of resources.
- Zoom out to align business goals with global goals and ally with the supply chain/value web to nurture circular, regenerative, and just experiences.

**Principles to follow to fulfil these aims:**

- **Values-led**

  o Guide the supply chain by example by connecting business purpose to greater global goals and ensure activities and output to nurture these goals

- **Responsible**

  o Implement the purpose within the business also to guide governance and foster an environment of empathy, diversity, and inclusion

- **Transparent**

  o Foster openness of activity, sharing of knowledge, and communication between the business, supply chain, and users to accelerate a more widespread transition to life-centredness

- **Accountable**

  o Use transparency and metrics to hold the business and governance accountable to life-centred commitments

- **Service-led**

  o Use service-based business models to convert mindsets of individual 'ownership' to 'access'

# 2.5
# Responsive
# approach

The three *Design Pillars* can be achieved by employing an attuned and responsive approach that takes a wider lens across space and time.

## Multi-level and multi-practice

Firstly, designing for both a human-centred product experience and a sustainable, regenerative, and fair lifecycle requires *zooming in and out* between view levels:

- **Micro level** — the individual user experience
- **Meso level** — the service or business model
- **Macro level** — the product's lifecycle impact at a global or national level

Changing view levels, attending to the needs of all stakeholders, and considering physical and digital product hybridisation requires merging and switching between multiple design practices

Example:

- Design for the target user with human-centred product design, inclusive design, and behavioural design.
- Then switch to systems thinking to view the lifecycle to maximise resource value, reduce waste, and ensure minimal impact on all peoples, non-humans, and planet.

## Time-honouring

Life-centred design also reminds us to consider the past, and to recognise alternate perspectives of the past, so that we may also consider the impacts of our design on future stakeholders.

- **Alternate pasts** — understanding the past of the problem and different perspectives of its history

- **Alternate futures** — speculating future scenarios to protect future peoples, non-humans, and planet

## Responsive and non-linear

Zooming in and out between levels, between human-centred design and lifecycle systems thinking, and from past to present to future, requires an adaptive approach that responds to projects and all stakeholder needs, whether its the target-user, the worker in a factory in another country, or the animals and environments impacted by supply chain.

This responsive approach enables life-centred designers to:

- Respond and adapt process to the needs of the interdependent stakeholders, from user to all peoples, animals, and planet
- Switch between design practices as they switch between local, organisational, and global levels
- Consider all perspectives of the past and possible future scenarios to protect all past wisdoms and future stakeholders

# 2.6

# The Life-centred Design Compass

*The Life-Centred Design Compass* is an illustrated guide to the scope and responsive process of life-centred design—a Responsive Approach to using the Key Practices to uphold the Design Pillars that respect all Interdependent stakeholders (Figure 17 - Life-centred Design Compass).

Please note, this diagram does not visually reference or align with the Doughnut Economy diagram.

The Compass aims to show that a life-centred approach is:

- **Human-centred**—At the centre is human-centred product design, with a focus on the target user and their use of the product. Life-centred design ensures this is also inclusive, responsible, and sustainable
- **Circular-focused**—Expanding product design to include product lifecycle thinking to maximise resource value, reduce waste, and regenerate all peoples, animals, and planet
- **Responsive in practice**—The arrows show the 'zooming in and out' between human-centred product design and lifecycle thinking:

  o From **inclusive** product design (accessibility and diversity) to **pluriversal** awareness and consideration (fostering and defending the thriving existence of many ways of being and knowing beyond the dominant globalised Euro-Western-centric one)
  o From **sustainable** use of materials and energy to **regenerative** initiatives in turn giving back to the sources
  o From **responsible** behavioural design to a business model **aligned** with global goals and supply chain partners

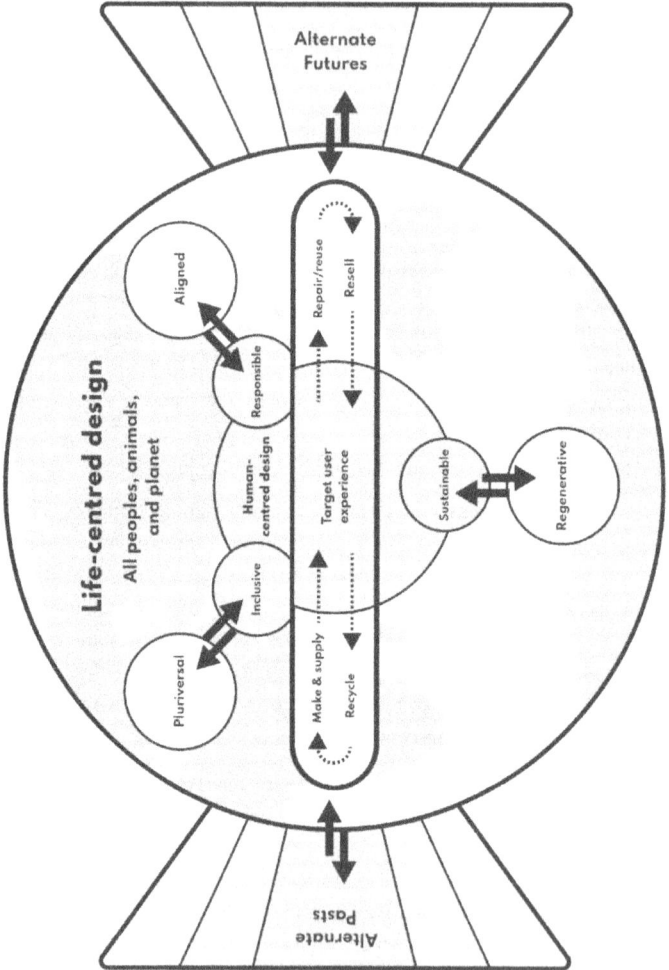

Figure 17 - Life-centred Design Compass

- **Time-honouring**—Arrows also show the switching between time views for considering impacts of design beyond the present:

  o **Alternate pasts**—understanding the past of the problem and different perspectives of its history
  o **Alternate futures**—speculating future scenarios to protect future peoples, animals, and planet

Every project would require different combinations of views and practices, methods which are still very much forming.

## What practices do I use and when?

As life-centred design is still very much merging, the strategies, tools, and combinations are also still evolving.

The key aspect of life-centred design, and what the Compass aims to express, is to use the various practices to zoom in and out and to check the impact of everything we design, to constantly balance the system view with the individual needs of all stakeholders past, present, and future.

Below are some examples of how different practices could be applied when focusing on different aspects of product/service design.

## Example 1— A product engineer designing a new product

With **human-centred design** and **inclusivity** at the core, a life-centred product engineer would remain aware of the pluriverse (**pluriversal design**) so their solutions don't perpetuate any misrepresentations or marginalisation from past or present; design for circularity (**circular design**); look to nature for design solutions (**biomimicry**); consider sustainable and ethical materials (**circular design**); design in ways to encourage sustainable and ethical user behaviour (**behavioural design**); and consider future scenarios of user behaviour and impacts to all stakeholders (**pluriversal thinking, interspecies design, and lifecycle thinking**).

## Example 2— A UX designer creating a digital experience

With **human-centred design** and **inclusivity** at the core, a life-centred UX designer would also remain aware of the pluriverse (**pluriversal design**) so their solutions don't perpetuate any misrepresentations or marginalisation from past or present; design in ways to encourage sustainable and ethical user behaviour (**behavioural design and sustainable digital design**); explore how to regenerate user digital communities (**circular design**); and consider future scenarios of user behaviour and impacts to all stakeholders (**pluriversal thinking, interspecies design, and lifecycle thinking**).

## Example 3— A service designer mapping a new service

With the target user in mind, a life-centred service designer might consider the full lifecycle (**circular design**) and understand the past of the problem to hear all peoples' perspectives (**pluriversal design and foresight**); look to nature for solutions (**biomimicry**); design circular (**circular design**); consider future impacts of system infrastructure and use on all stakeholders (**pluriversal thinking, interspecies design, and lifecycle thinking**).

## Example 4 — A business looking to become more life-centred

An existing business looking to become more life-centred might engage service designers and supply chain partners to identify leverage points for circular, inclusive, and pluriversal improvements on their product, business, culture, and supply chain (**pluriversal thinking, interspecies design, lifecycle thinking, biomimicry, systems thinking, distributed design**); identify opportunities for regenerating peoples, sentients, and planet along the lifecycle; reduce the ecological footprint of the digital channels (**sustainable digital design**); and speculate future impacts of their proposed life-centred changes (**pluriversal thinking, interspecies design, and lifecycle thinking**).

You can also explore these examples in this interactive Compass tool:

Compass viewer

Designers of physical and digital products can use *The Life-centred Design Compass* as a learning tool to understand the breadth and complexity of life-centred design.

Embracing this complexity will help product designers develop skills in multiple practices which they can then apply to greater initiatives by designing a healthier and kinder world.

# Part
# 3
# -
# The Practice

# Life-centred design strategies and lenses

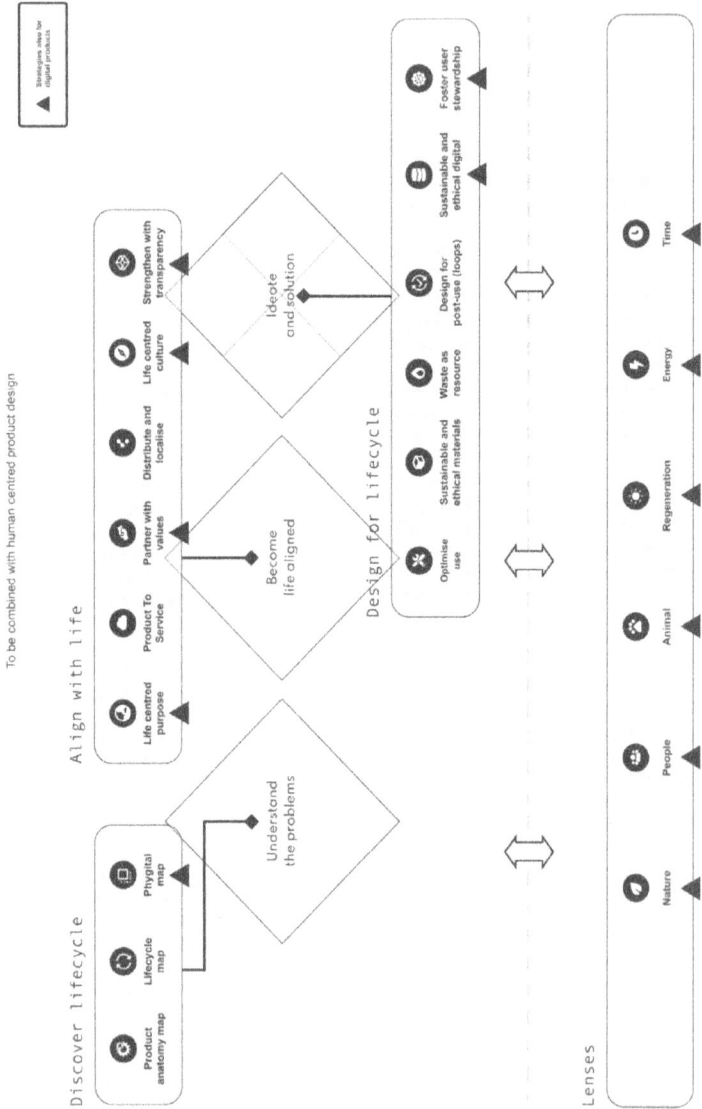

To be combined with human centred product design

Strategies view for digital products

**Discover lifecycle**
- Product anatomy map
- Lifecycle map
- Phygital map

Understand the problems

**Align with life**
- Life centred purpose
- Product To Service
- Partner with values
- Distribute and localise
- Life centred culture
- Strengthen with transparency

Become life aligned

Ideate and solution

**Design for lifecycle**
- Optimise use
- Sustainable and ethical materials
- Waste as resource
- Design for post-use (loops)
- Sustainable and ethical digital
- Foster user stewardship

**Lenses**
- Nature
- People
- Animal
- Regeneration
- Energy
- Time

Figure 18 - Life-centred strategies and lenses

172

# 3.1
# Introduction

Keeping human-centred product design at the core, and using circular strategies as a base, life-centred design infuses both practices with the principles of the other nine practices. This fusion of practices produces the life-centred strategies, lenses, and methods.

## Strategies

Strategies can be used at specific stages of the design process:

- Discover lifecycle—understand the lifecycle of the product/service and its impacts on all stakeholders
- Align with life—align the business model with global goals
- Design for lifecycle—optimise the life-centredness at every stage

Keep in mind these strategies work *in conjunction with human-centred design* and can be implemented into variations of the Double Diamond. A triple diamond example can be seen in Figure 18 - Life-centred strategies and lenses.

## Lenses

*Lenses* can be used at any stage and at any level to view the process and solution through key considerations to optimise life-centredness.

## Methods

Methods are combinations of Strategies and Lenses.

**Please note:**

- While some of these strategies are from the earlier exercises they are repeated here for completeness
- On (FIGURE), the inverted triangles denote strategies and lenses that be applied to both physical and digital products

# 3.2
# Strategies

## 3.2.1 Discover lifecycle

These strategies can be used to understand the lifecycle of the product/service and its impacts on all stakeholders

- Product Anatomy Map
- Lifecycle Map
- Phygital Map

## Product Anatomy Map

*Tool: Product Anatomy Map*

*Download from lifecentred.design*

The *Product Anatomy Map* (Figure 19 - Product Anatomy Map) is a simplified product *Disassembly Map* to map the product's parts (and packaging) to help you to:

- Get to know the anatomy of a product
- Grow awareness of the resources and energy required for a product
- Practise with life-centred *Strategies, lenses, and methods*

Use the *Product Anatomy Map* to record the parts, the order in which they are removed, the fasteners/adhesives used, the tools needed, difficulties, key components, and the time it takes.

- **Find a discarded product**

  o Practice with an unused product or source one from waste, such as discarded items left in rubbish rooms or on the street. Perhaps set yourself the purpose of accessing a broken product to map how easy or hard it is to access the part that needs repair.

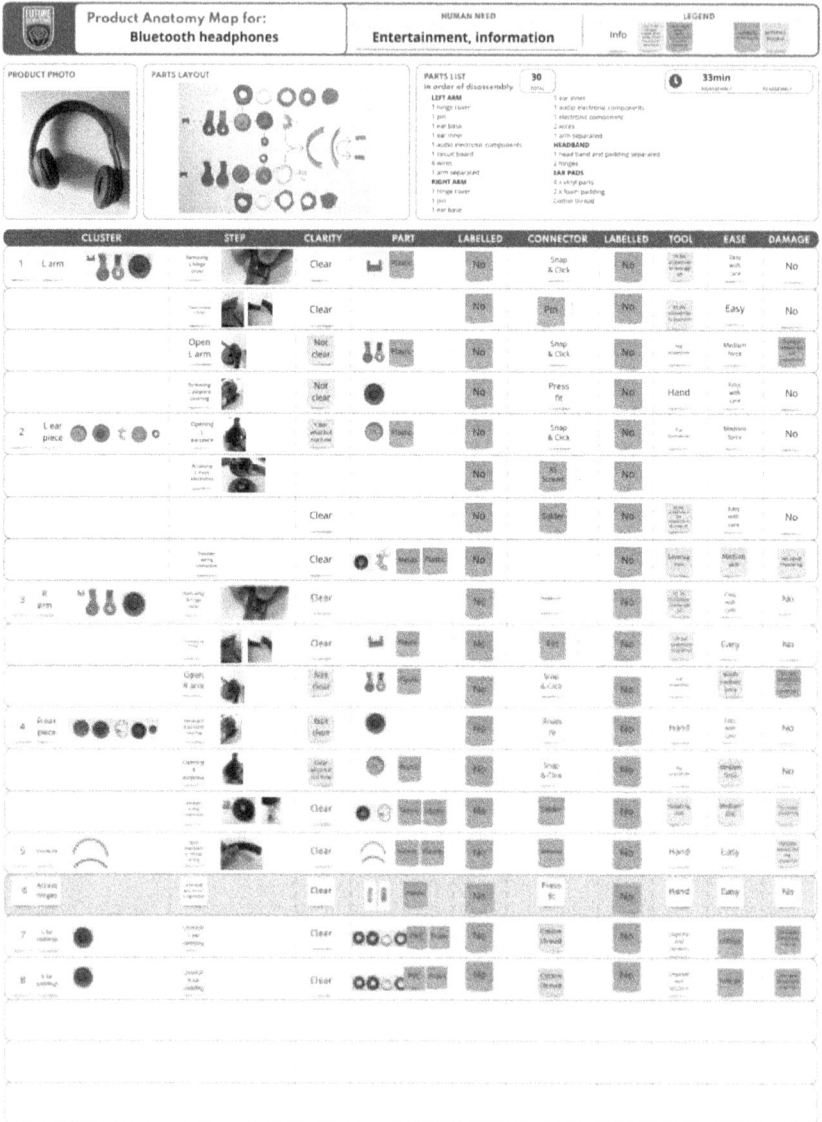

Figure 19 - Product Anatomy Map

- **Preparation**—To map disassembly, you'll need[150]:

  o Room to layout all parts
  o Tools (screwdrivers, pliers, wire cutters, etc.)
  o Measuring tape or ruler
  o Note capturing materials (pen, sticky notes, camera, an online whiteboard like Miro, Mural, etc.)
  o Depending on the complexity of the product, you might need containers of some sort to collect fasteners, etc. for each step (bowls, drinking glasses, etc.)
  o Depending on the size of the fasteners, etc., you might also need tweezers and a magnifying tool
  o Protection cloth, etc. for delicate components
  o Refer to manufacturer resources for guidance
  o Refer to *ifixit.com* for help and developing knowledge

- **Disassemble**—As you disassemble the product and layout all the parts, use the *Product Anatomy Map* to record[151]:

  o Clusters of parts
  o Each step to disassemble
  o Clarity of disassembly—clarity of what to do, visibility/identifiability/accessibility of connectors
  o Individual parts and connectors used (fasteners and adhesives) in order of disassembly, the basic material type of each part (e.g., wood, metal, plastic, coating, pigment, stone, engineered stone, foam, cotton, wool, nylon, etc.), and if the materials are labelled
  o Tools required at each step, and if they're standard or specialised
  o Ease or difficulty of manipulation
  o Any damage caused during disassembly
  o A diagram flow of the parts and fasteners removed in order
  o Key components and where they are in the disassembly process:

    ▪ Failure—highest failure rate or functional importance
    ▪ Environmental—the most environmentally harmful
    ▪ Economic—the components with the highest embedded economic value
    ▪ Recovery/reuse—high potential for reuse/recovery[152]

  o Total number of parts
  o Total list of parts in order of disassembly
  o Time it takes to disassemble and reassemble

From this mapping, you can refer to the *Design for lifecycle* strategies to identify where to make life-centred improvements.

For example, the mapping in Figure 19 - Product Anatomy Map highlights the problem of a part that is most likely to break (the hinge) being one of the most inaccessible parts. This product design could benefit from a review of the *Optimise use* strategy. Or you can then use the *Sustainable and ethical materials* strategy to explore the materials in more detail and choose new ones that are less toxic and/or stay in loops longer.

If your product is 'purely' digital, use the *Phygital Map* tool to discover the physical components supporting the delivery of the product/service.

# Lifecycle Map

**Map the resources and activities supporting the end-to-end product lifecycle to understand the peoples, animals, and environments impacted**

*Tool: Lifecycle Map*

*Download from lifecentred.design*

Designing to keep products in circular loops of reuse is about:

- Maximising the materials in use and extending their value to keep more raw materials in the ground (this includes maintaining materials' inherent properties untainted for as long as possible)
- Minimising waste and extraction of new materials

Use the Lifecycle Map (Figure 20 - Lifecycle Map) to map the flow of materials and parts of your product across its full lifecycle to get to know their impact on people, animals, planet and then start brainstorming life-centred interventions.

The *Lifecycle Map* is best used after mapping your product's parts/materials, but you can also use it to discover lifecycle thinking.

Start by mapping one part/material—the main material, the most expensive or most processed, etc. Then also include all other materials to create a visual representation of the lifecycle ecosystem of your product and its relationships—what it takes from people, animals, and planet, and the value it gives.

From this mapping, you can refer to the *Design for lifecycle* strategies to identify where to make life-centred improvements.

The Lifecycle Map can be used to:

- Audit material use and flow to identify improvement opportunities
- Audit the environmental, social, and economic impacts to identify opportunities for life-centred improvements
- Map an ideal future 'Ambition' state for strategizing
- Initiate lifecycle thinking

If your product is 'purely' digital, use the *Phygital Map* tool to discover the physical components supporting the delivery of the product/service.

## The Lifecycle Map explained

The header row is where you write your product name, tick whether it is a 'Current state' audit (as it is today) or an 'Ambition state' (an ideal version to aim for) and write the key **HUMAN NEED** that the product fulfils.

The **LEGEND** suggests how to use colourised sticky notes to capture the materials, energy, impacts on people, planet, and finance, and ideas for life-centred improvements. The idea of this colour palette is that the more your life-centred strategies reduce non-renewable energy and negative impacts (red and orange sticky notes), the greener and more blue your map becomes—so you visually 'cool' the lifecycle as you 'cool the planet'. The tool has two sections:

- **RESOURCES**—the technological and biological resources used to create and maintain your product
- **IMPACTS**—the impacts of the product lifecycle on people, animals, planet, and business financials.

The **RESOURCES** section is where you capture the materials and energy used to create your product and how they flow through the lifecycle.

The flow starts in the left-hand column 'Material extraction', and flows into the **CIRCULARITY** section—this is the supply chain and product use flow, that then loops back into circular loops of parts/material reuse:

- The top resources row is the **Technology** row, to record materials that are processed beyond their raw state (metals, plastics, pigments, etc.)
- The bottom row is the **Biological** row, for materials kept in their natural state (cotton, wood, etc.)
- The middle row is for recording the energy used or created at each step (also capture here the energy used/created during the loops)

The **IMPACTS** section is where you capture the positive and negative impacts on people, animals, planet, and finances, at every stage—these stages align with the same ones in the **RESOURCES** section above.

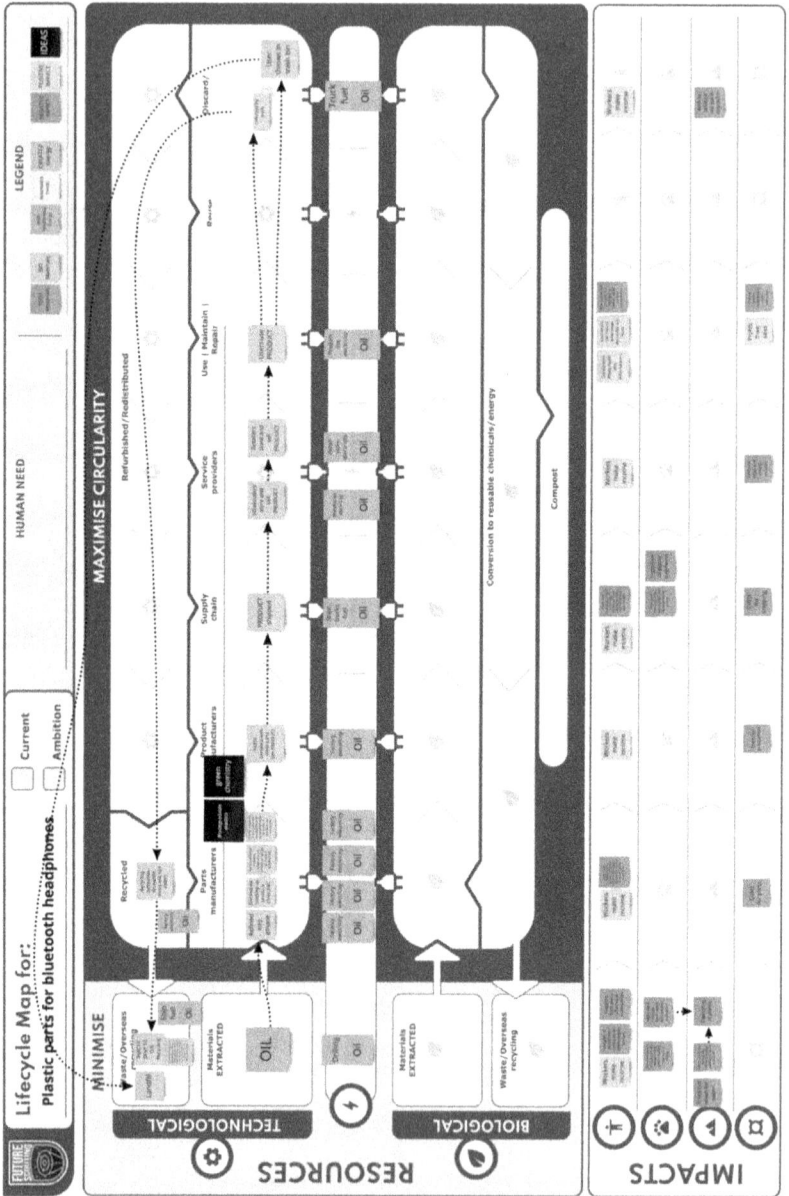

Figure 20 - Lifecycle Map

**Mapping the resources**

In the **RESOURCES** section, start at the Materials Extraction column, and use the materials sticky notes to:

- Capture any materials of your product that you can identify, and place them in the technological or biological boxes
- Map these materials through the entire journey, noting what happens to them—if they are transformed into new materials, attached to other parts, separated at collection or recycling, etc.
- You can create animal personas for the key environmental elements/resources to learn about them (and to keep them in mind through the design process)
- Use research and data to inform your mapping

**Next, using the energy column and the energy sticky notes:**

- Map the energy used or produced by what happens to the material/part at each stage.
- Consider if any energy (electrical, heat, kinetic, etc.) produced isn't being used

**Mapping the impacts**

- In the **IMPACTS** section, starting at the left-hand side, consider the life-centred stakeholders in the first 3 rows (people, animals, planet) and economics in the bottom row (cost and profit), and map the negative and positive impacts:

  o All peoples

    - Target users
    - Non-users—Individuals, communities, and employees of organisations working within the product lifecycle
    - Invisible humans—individuals and communities not involved in the lifecycle but who are impacted by it
    - All human knowledge and ways of existing
    - *Create human personas for the key humans*

  o All animals

    - From large animals (amphibians, reptiles, birds, and mammals) to insects and microbes; on land, sea, air, or underground; domestic, livestock, captive, or wild; whether 'proven' sentient or not
    - *Create animal personas for the key animals*

  o All planet

- Vegetation (trees, forests, swamps, etc.)
- Water systems (oceans, lakes, rivers, freshwater)
- Air
- Soil
- Climate and weather
- Landforms (mountains, hills, etc.)
- Sunlight
- Noise
- Temperature
- Gases and atmospheric elements
- Biodiversity
- *Create environment personas for the key environmental elements*

  o Financials

  - Costs and profits to the business

Fill any knowledge gaps with research and data, then assess to determine the best opportunities for improvement and innovation.

**Assess the current state:**

- How much red and orange do you see? This is an immediate visual indication of how life-centred your product is
- How many materials/parts stay inside the **CIRCULARITY** section, and how many pass outside the loops and end up as waste? Can any of these waste materials be replaced to keep them within **CIRCULARITY**, or can they be converted to a resource/fuel for something else?
- Are there concentrations of negative impacts to focus on?
- Are there concentrations of positive impacts to enhance?
- Use the *Design for lifecycle* strategies to prompt assessment questions and points of intervention, and capture ideas using the black opportunity sticky notes

**Choose a strategy**

With opportunity points identified, prioritise them for further exploration by using a 2x2 grid (E.g., 'impact' versus 'effort'—use the *Gather & Prioritise tool*).

**View the lifecycle through Lenses**

Use the *Lenses* to view the impacts through different perspectives, and capture problems and ideas using the black opportunity sticky notes.

For example, you could use a *Life-centred Design Workshop* from the *People Lens* to explore one lifecycle stage to better understand impacts to the people and place or view your design decisions through the *Time Lens* to assess their impacts on the future.

**Create an Ambition Lifecycle map**

Dream big with an ambition lifecycle map, go for gold standard, and use it to create a future vision for strategizing.

# Phygital Map

**Map the physical components supporting the end-to-end digital product lifecycle to understand the peoples, animals, and environments impacted.**

*Tool:   Phygitial Map*

*Download from lifecentred.design*

Digital products require physical components to support their existence—such as servers for data processing, retail outlets, etc.—and they drive need and use of physical devices and services. A photo editing app, for example, require servers and energy derived from physical energy generators. Uber is supported by servers, energy, a smart phone to access it, smart phones for drivers, and it generates the use of vehicles and roads.

The *Phygital Map* (Figure 21 - Phygital Map) is a tool to aid in mapping a high-level view of these physical resources supporting a digital experience. It isn't a detailed or decision-making tool, it's a thought initiator to identify more detailed lines of investigation.

You might identify resources that you have direct influence over—you can map some of these through their own *Lifecycle Maps* to identify opportunities for life-centred improvements. And for what you don't have direct influence over, you might be inspired to regenerate them indirectly.

## The Phygital Map layout explained

The map is a time flow of columns from left to right:

- User finds and gets—how the user finds and purchases access to the digital product (sales channels, website, ecommerce, delivery, etc.)
- In use—how the product is used by the user, and any support required during that time
- End of use—how the use of the product ends, what happens to the user's data, etc.
- Maintenance—any other operational activities to keep the digital product active

The rows capture actions made by User, Provider, and 3rd parties:

- User:

  o Digital—the actions of the user in the digital world during their digital experience

  o Physical—the correlating physical components that enable user's experience (device, modem, etc.)

- Provider (Product owner/service provider):

  o Digital—any digital experience created/enacted by the provider to enable the user's actions (website, processing online purchase, etc.)

  o Physical—the correlating physical components that enable the provider's actions (servers, etc.)

- 3rd Party (any 3rd party service providers required to support the provider's or user's actions):

  o Digital— any digital experience created/enacted by a 3rd Party to enable the provider's or user's actions (cloud service, PayPal, etc.)

  o Physical—the correlating physical components that enable the 3rd Party's digital actions (servers, etc.)

- This should provide a high-level view of the overall physical component of a digital product's experience. Look at the map through *Lenses*:

  o How are *People* affected?

  o How will continued use impact the physical resources and people over *Time*?

  o Can any components be transitioned to being powered by renewable *Energy*?

  o Are there *Relationships* that can be leveraged to explore and foster life-centred collaboration?

- Identify any physical and digital components you have direct influence over:

  o For physical components, take them through a *Lifecycle Map* to identify *Design for lifecycle* strategies to improve their life-centredness

  o For digital components, use the Sustainable and ethical digital strategy in Design for lifecycle

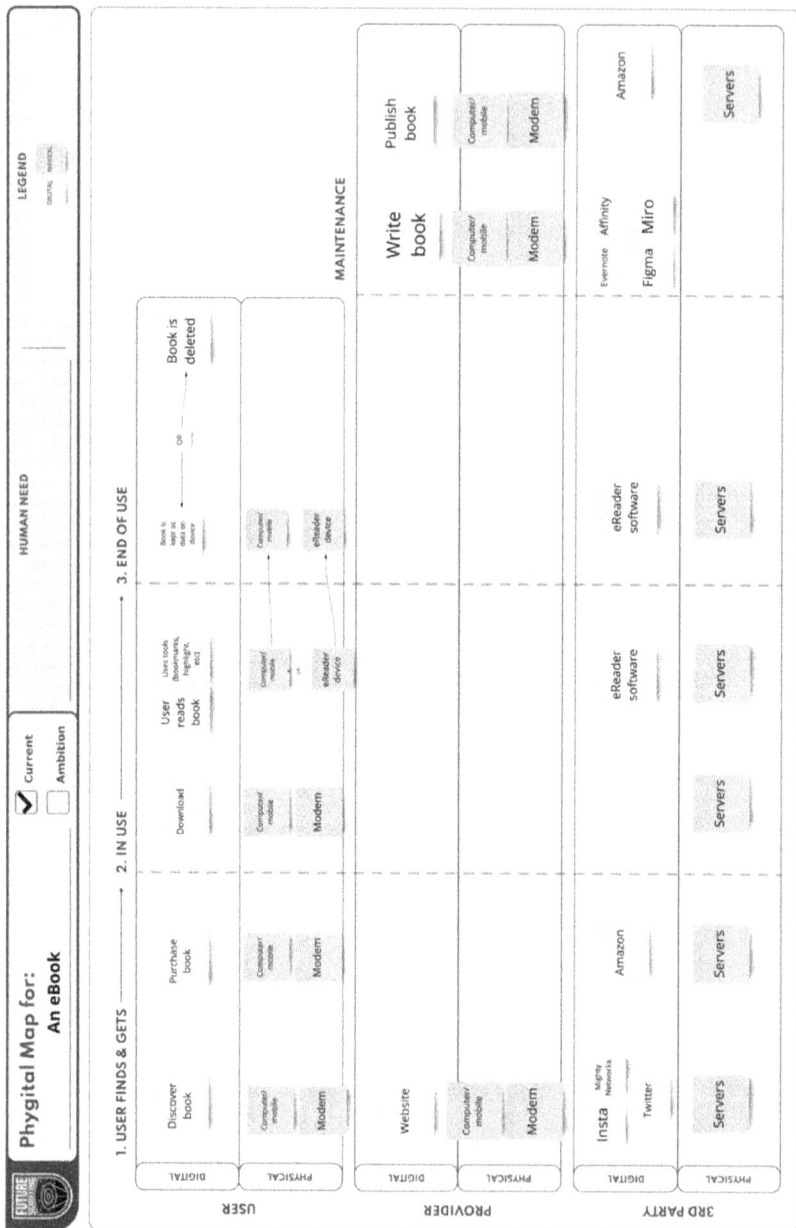

Figure 21 - Phygital Map

- Look at all the physical components again, including the ones you can't influence, and assess for commonalities in resources for opportunities for Regeneration:
  - What are the main devices used?
  - What are the main materials used in these devices?
  - What planetary resources do these take from?

Look at this through the *Regeneration Lens* to explore how your business can give back to these resources.

# 3.2.2. Align with life

These strategies can be used to align the business model with global goals:

- Life-centred purpose
- Strengthen with transparency
- Convert product to service
- Partner with values
- Distribute and localise
- Life-centred culture

# Life-centred purpose

*Tool—Life-centred Purpose*

*Download from lifecentred.design*

**Foster regenerative synergy can be fostered by aligning a business purpose with global goals.**

As discussed earlier, a regenerative business goes beyond sustainability and shares prosperity with the wider ecosystem of the product lifecycle to restore, renew, and revitalise every connected partner, community, and environment, nurturing resilience for all and generating innovation.

A life-centred purpose should be succinct about what to do and what not to do and grounded in possibility, yet inspiring in its potential. It should guide everything from strategy and operations to culture[153], with diversity, inclusion, and fairness as defaults, and utilising regenerative concepts to elevate and thrive the potentials of individuals[154]. Finally, a life-centred purpose should hold its purpose accountable by establishing measures, timelines, and deadlines to safeguard real progress and earn credibility.

## Using the Life-centred Purpose tool

Use the *Life-centred Purpose* tool (Figure 22 - Life-centred Purpose) to explore aligning the purpose of your product, business, or personal projects with global goals to initiate a transition toward more life-centredness.

## Page 1

- Complete the header banner:
  - o Write the subject (business, product, or project name) at the top
  - o Write the value the subject adds to its customers lives, or the value your personal work provides

- In the 'PEOPLES/NON-HUMANS/ECOSYSTEMS' panel, note who and what is impacted by your product/project's lifecycle (refer to your Lifecycle Map from Exercise 1)
- Rewrite these problems as *How Might We?* statements to foster innovation thinking
- Brainstorm ways your product/project could alter its lifecycle or give back to these peoples, animals, and/or planetary resources.

  o Nurture:
    - All life-centred stakeholders (people, animals, and planet) who might be impacted by your product's lifecycle
    - At all levels—world, organisations/communities, and individuals

  o Think in terms of:
    - Product and service innovations
    - Business model innovations
    - Sustainable and inclusive opportunities within your organisation

  o To brainstorm further:
    - Use the QR codes at the top right of the page for ideas
    - Research ideas
    - Use your *Ambition Lifecycle Map* as guidance
    - For physical products, create a *Product Anatomy Map* and/or *Lifecycle Map* and then explore life-centred strategies
    - For digital products, create a *Phygital Map* to identify the main physical aspects of your digital that you can directly influence or indirectly regenerate

  o Think big and small, you can prioritise later

- Review the global goals at the bottom of the page and tick all that your product/project could contribute to, keeping in mind your product/project's value-add and the ideas you just brainstormed (scan the QR code for more detail on the goals). Try to connect with at least one goal from each of the three groups (social, environmental, and governance), but just do what feels right and makes sense
- Brainstorm more ideas about how your product/project could contribute to these goals and the ideas you've already generated

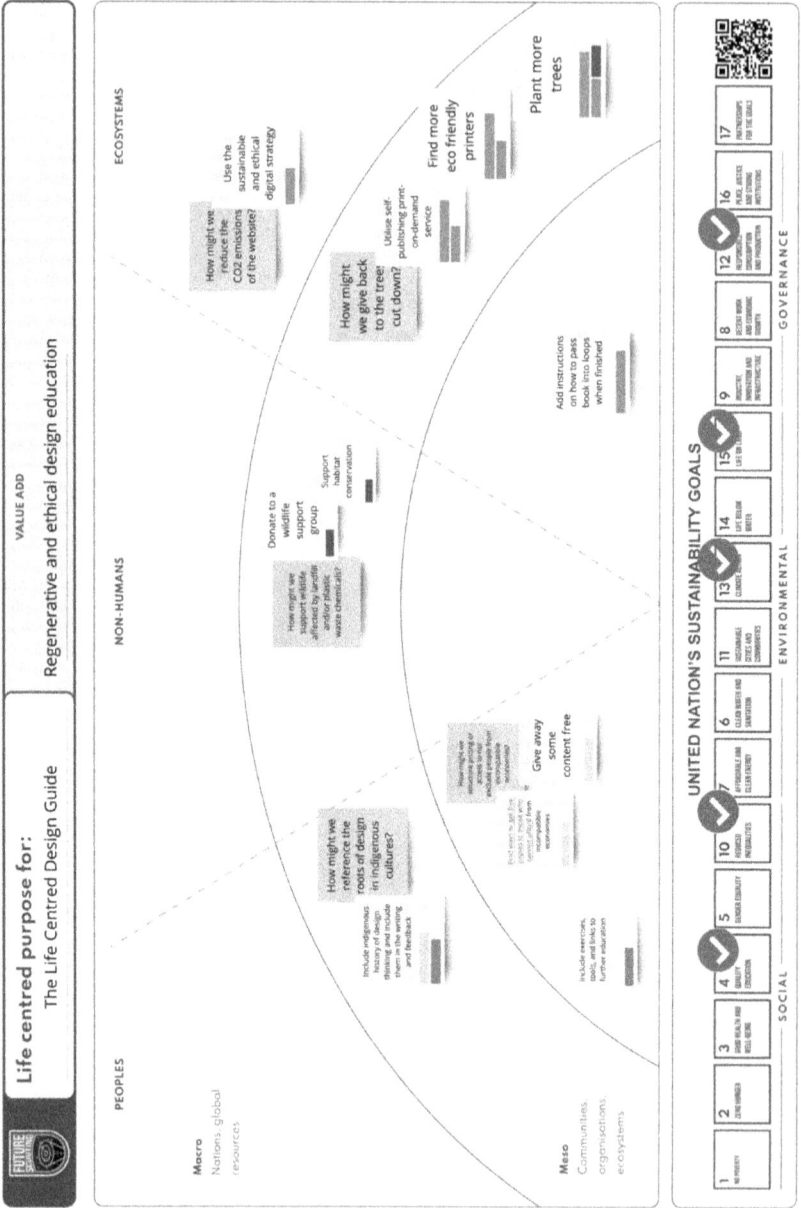

Figure 22 - Life-centred Purpose

**Page 2**

- Sort all your ideas and start merging and refining them. Explore them in a bit more detail and refine them into definite actions.
- Use the FEASABILITY grid on the right of the page to identify the best goals to explore further (the ones that land in the shaded quarter have the most positive impact and are ones you have the highest ability to accomplish). You might also narrow your goals down to a few
- Brainstorm ways to maintain accountability (metrics, deadlines, etc.)

**Page 3**

- Following the prompts, write a life-centred purpose statement to include these goals and metrics. Note your global goals at the bottom for future reference.

**Alternate global goals**

In the spirit of including all voices and perspectives, the Life-centred purpose tool includes a version with the values defined in Pluriverse: A post-development dictionary as an alternative set of global goals.

**My example**

I used this tool to establish a more life-centred purpose for this guidebook that extended its impact beyond just sharing information about life centred design. You can view this example at lifecentred.design.

# Convert product to service

Tool—Product to Service

*Download from lifecentred.design*

**Explore how you can transform your product idea into a service-first approach using business models such as renting, subscribing, leasing, sharing, or on-demand access.**

### Using the Product To Service tool (Figure 23 - Product to service):

- Complete the header banner:
    - o Write your business name at the top
    - o Write the value the business brings to customers

- Choose a service model to explore:
    - o Renting
    - o Subscribing
    - o On-demand access
    - o Leasing
    - o Sharing
    - o Other

- Brainstorm what the service experience might involve:
    - o This simplified experience blueprint starts at the left and breaks the journey into **steps**:
        - ▪ USER FINDS & GETS—How would the user first find/engage with the business and initiate the service contract?
        - ▪ IN USE—After signing up, how would the user begin using the service, what might the onboarding be like, what happens if the product fails, etc.?
        - ▪ REPAIR—How would the product be repaired during use? Would it be returned to service provider, or are tools included for user-repair? Are online repair workshops held, set up by the provider or a community?
        - ▪ END OF USE/RETURN— How would the user return the product or end the service? How would the product be collected?
        - ▪ NEXT USE—What business activities would be needed to prepare the product for the next user, any maintenance refurbishing, etc.?
    - o The rows view the **steps** through different **lenses**:

- USER EXPERIENCE—What the user would experience, need, say, or do?
- BEHIND THE SCENES—Any business system/operation working in the background to facilitate what the user is doing (include any reference to 3rd parties)
- STAFF ROLES—What staff might do, need, or say, and any new roles required?

- Benefits
  - After exploring the service experience, list the benefits the model might offer the user and the business.

- Transition Milestones
  - As an exercise to start envisioning the transition to a service model:
    - Brainstorm the key milestones that might need to be reached to complete that transition
    - Think about the order tin which they would need to happen

Complete this worksheet for as many service models as might suit your offering and compare benefits versus effort.

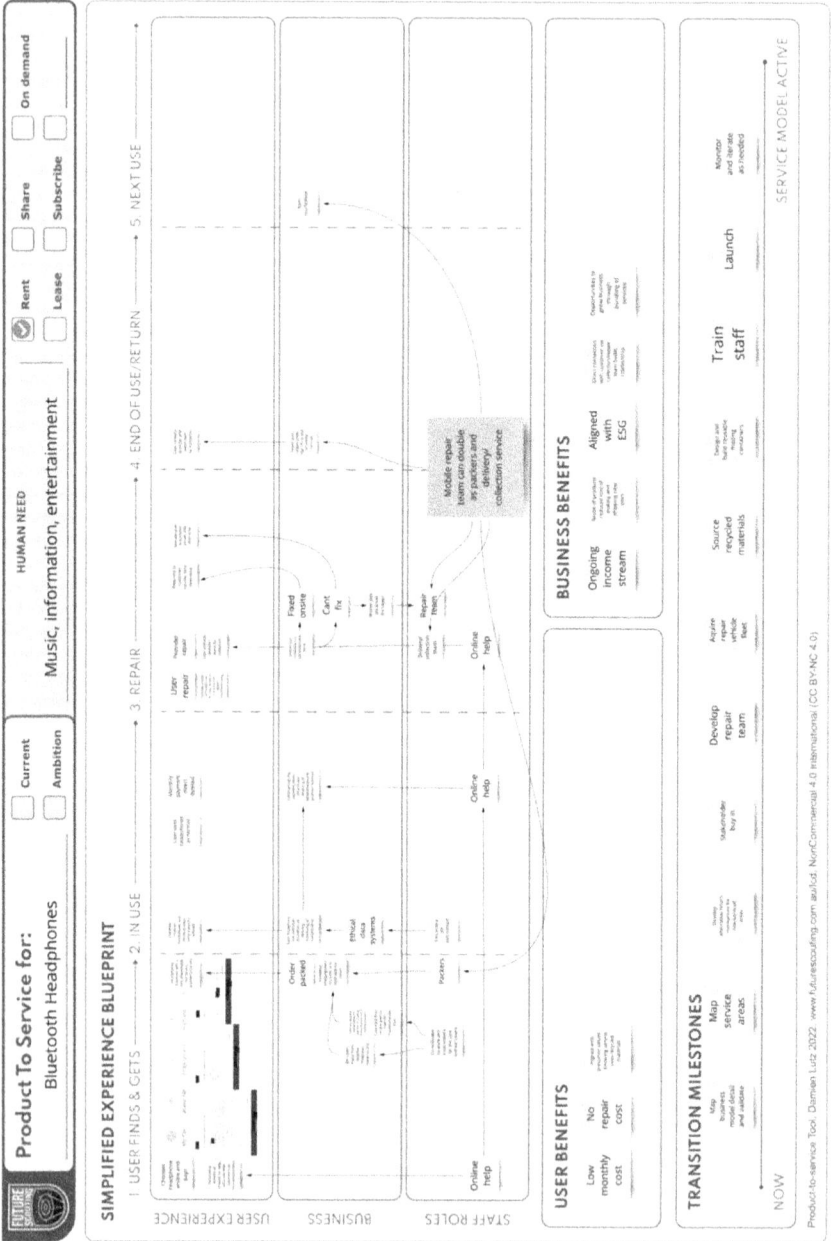

Figure 23 - Product to service

# Partner with value

**Improve the circularity of your product by partnering with the value web to share and maximise value for all.**

## Value web

Rethink the supply chain as a value web and leverage partnerships from the lifecycle to improve the sustainability of the service ecosystem.

### Get to know your partners

- Create a map of partner locations to visualise the breadth of your product's lifecycle ecosystem, and to better understand their values:

  - What's their business model?
  - Are they already circular—do they receive product, parts, or materials back for repair, refurbishment, or recycle, etc?
  - What are their incentives?
  - What power and influence do they have?
  - Who or what has power and influence over them?
  - What languages do they use?
  - What technology do they use?

- Repeat the process for all partners

### Look for points of connection

Map a systems view of all the partners to better understand the connections and overlaps (see the *Regeneration Lens*):

- Imagine the supply chain is no longer linear, but circular, and map an ideal state
- Look for leverage points such as overlaps of activity, common resources, common journeys, etc.—how could the resources of all the value web partners be reconfigured and optimised? For example, can the transporters also collect the product at the end of use to return it to the part or product manufacturers for refurbishment, repair, or recycling?
- Use the *Gather and Prioritise* tool to find themes and starting points

### Connect and collaborate

As a means of initiating connection, organise a meeting with partners to explore potential collaboration. Share your circular value web ideas or any of your own life-centred design efforts to lead by example:

- Run a *Life-centred Workshop (see the People Lens)* to explore ESG commitments and collaboration that might be mutually beneficial

- Invite them to internal design workshops to allow them to contribute their expertise. Include distribution and ecosystem actors in design sessions for practical insight

**Also consider these types of partners:**

- Brand
- Financial
- Affiliates
- Additional value-add products

# Distribute and Localise

**Consider ways to use an open-source mentality and moving data instead of product to distribute and localise the design and production process.**

Prompts

- Start with the main areas/regions where your sales are greatest, and research what nodes of maker shops/spaces or fablabs that can produce supply on demand
- Establish an open and collaborative sharing of ideation, prototyping and testing between headquarters and nodes
- Run life-centred design jams to explore nurturing the pluriverse and connection with place by allowing each region to localise your product through other locale-specific needs, materials, rituals, cultural symbols, and/or visual treatments
- Try *Deepa Butoliya's* decolonising and localising material mapping exercise with your nodes[155]:

  o Dismantle the product and consider who and what are impacted by all the separate parts
  o Reconstruct the product using only resources from your local areas, to 'decolonizes the immediacy and materialism of modern design[156]'

# Life-centred culture

**Life-centre the business culture by aligning governance, ownership, finance, and workforce diversity with the life-centred purpose.**

## Governance, ownership, and finance

**Prompts:**

- Who is included in discovery and decision-making?
- Who and how are progress metrics monitored?
- Is life-centred cost accounting implemented that considers 'cost' to humans and the environment[157]?
- Who has ownership of the organisation and its assets (digital and physical) and what other organisations are they affiliated with that might affect their values?
- Who funds the organisation and what other organisations are they affiliated with, that might affect their values?

## Workplace diversity

Use existing resources and team-building exercises, workshops, and safe spaces for open discussions to listen and discover barriers:

- Flip barriers into opportunities to improve
- Create policies and guides for education, training, and development (e.g., for language use and decision-making)
- Track for continuous improvement
- Search for talent in non-traditional areas, connect with universities, youth groups, etc.
- Look to embed a framework like DEIJ in the business (Diversity, Equality, Inclusion, Justice)—*jedicollaborative.com/pathways-action/self-guided-pathway*

# 3.2.3. Design for lifecycle

These strategies can be used to optimise the life-centredness of a product's construction, use, and reuse.

- Optimise use
- Sustainable and ethical materials
- Waste as resource
- Design for post-use loops
- Sustainable and ethical digital
- Foster user stewardship

# Optimise use

**Maximise the usable life of your product, and design for easy and clean disassembly to enable repair, recovery, refurbishment, up-cycling, and recycling of parts and materials.**

With an overall view of your current state, you might want to improve the product design itself to extend the sustainability of first uses.

'First uses' can be defined as the use after first purchase and first use by anyone is sold onto without the product being disassembled or entering circular loops.

### Durability

Designing in durability is key to circularity, as durable products stay in first use longer, delaying the inevitable degradation of materials through refurbishment and recycling loops. Products that retain integrity and aesthetics also increase chances of resale when the first user is finished with the product, which also extends their value by delaying them entering the loops.

**Prompts:**

- What is the average lifespan of use, and can it be extended? What are the number of lifecycles you're designing for your product (how many times can it be repaired and refurbished before materials breakdown)
- Can you use more durable Sustainable and ethical materials?
- Can you use different structural techniques to strengthen the product?
- Can you blend materials differently?
- Are there any activities your customer/user can perform to sustain its life and reduce the need for repair? Does your product need a 'grease and oil' occasionally?

- Can you Digitise the product to monitor and maintain itself in anyway? Can you encourage users to set up reminders to encourage maintenance?
- Review your instruction/care manual—is it easy to read and understand? Does it support multiple languages? Is it easy to keep nearby? Can you add a QR code to your product so the user can easily retrieve online information?
- How can you connect the user emotionally with the product to foster more care and use extension? Use storytelling to foster the valuing of older, long-lasting products over new

## Modularity

Designing for modularity subdivides a product into modules that work as a whole, but which can be created, modified, and/or replaced independently—like LEGO—while remaining standardised for compatibility of replacement parts, etc. Modularity allows products to be adapted by the user to accommodate upgrades and different contexts that change with environment or time. This makes personalisation, upgrading, repairing, and end-of-life options use less energy and resources and produce less waste.

Adaptability is important not just for differing abilities, but to allow for customisation according to the needs and values of different places—inclusive design should also foster the thriving of diversity, not just satisfy a specific problem or the minimum needs.

### Prompts:

Look at your Product Anatomy Map and consider:

- What parts need to be replaced at the earliest, and are they easy to access? Can you design them in a way to be removed without disturbing the product's overall integrity?
- Does your user's needs or methods of use change over time? For example, does the product need to adapt to growing bodies or changing abilities?
- Aim for designs that are as adaptive and multi-modal as possible to benefit many

## Repairability

Products designed for repair greatly extend their life by keeping them in the early circular loops. Make sure both user and manufacturer can repair with access to instructions, spare parts, and the systems that enable and encourage repair culture.

Look at your Product Anatomy Map and consider the following:

- Note which are the most critical components and the ones needing repair the most often the most accessible? Consider what you would do with these parts to keep them out of landfill. Note the effort and energy it would take to do so, and what parts would still end up in landfill.

- Ensure any repair tutorials you create are easy to read and understand. Do they support multiple languages? Are they easy to keep nearby? Can you add a QR code to your product so the user can easily access online information?
- Are the tools and parts required for repair affordable and easily accessible by both user and manufacturer? Can you supply 3D print files for parts?
- Can the product monitor and repair itself in anyway?
- Ensure repair tutorials and relative support continue for discounted models

## Energy efficiency

The energy products require during their use is a major source of environmental impact. Consider both the source of your product's energy and the amount required to operate.

- Use renewables energies for use or part use (e.g., solar powered, kinetic energy) and recharge (rechargeable batteries)
- Use timers for automatic switching to power save mode and deactivation
- Research other energy saving techniques specific to your product and technologies
- Should its battery be charged before its power reduces below a certain percent to maximise efficiency?

# Sustainable and ethical materials

Choosing better materials for your product and packaging can be complex and require detailed material identification and expert assistance. But anyone can start learning by going beyond their realm of responsibility or influence to know about their product's impact on the world.

**Considerations for life-centred material choices:**

- Sustainable materials:

  o Recycled
  o Supportive of disassembly and repairability by being easily separated at end of use for recycling
  o Biodegradable, like natural fibres (cotton, wool, timber, etc.), or lab-grown biomaterials (plant-based and lab-grown materials, like mushroom leather)
  o Have extraction methods with low carbon footprint

- o Lighter materials might mean less energy required to transport and less materials needed for packaging
- o Are any of these materials consumed faster than they are replenished, or are they already running out? Replace with more sustainable materials, like those certified for being responsibly managed for environmental, social, and economic benefits
- o Using materials 'honestly' and never making them seem different to what they are to reduce the processes, energy, and material waste, material contamination, and disassembly hindrance[158]

- Safe materials:

  - o Non-toxic materials, avoiding chemicals like those advised by *materialwise.org or the* Cradle to Cradle Certified™ Banned List of Chemicals

- Ethical and just materials:

  - o Materials that do not originate from areas known for conflict materials (the extracting and selling of natural resources from a conflict zone that perpetuate the conflict
  - o Materials that do not originate from areas or organisations known for forced and/or child labour
  - o Look for the material suppliers' names and web search for information on their material partners and countries of origin, and research if these are known as areas of conflict materials, forced and/or child labour, etc.
  - o Read the Ethical Sourcing Toolkit from Positive[159]

You can also aim for Cradle to Cradle® certification, a world standard for safe, circular and responsible material use that assesses materials across five categories[160]:

- Healthy materials that are safe for humans and the environment
- Product Circularity
- Clean Air & Climate Protection
- Water & Soil Stewardship
- Respecting human rights and contributing to a fair and equitable society

Requesting a manufacturer to look for alternative sources or sharing with them a 'Restricted Substances List' might be simpler methods for transitioning to more sustainable and ethical materials, but there is a risk when not defining what can or should be used resulting in ineffective or worse substitution[161].

## Dematerialise

Reduce the resource requirements of your designs by minimising the number of different materials needed for the product and its packaging.

- Design with *materials* that can reused throughout the design
- Design *components* that can be reused throughout the design
- Redesign components to use less material
- Use lighter materials
- Eliminate unnecessary packaging:

    o Is the packaging necessary at all? Can it be done with less materials?
    o Use only one material if possible
    o Use alternative securing methods to adhesives if possible
    o Make it compostable or recyclable
    o Repurpose packaging to avoid single use
    o Incorporate the packaging into the product
    o Give it an informative aspect (e.g., educating the user)
    o Make it compostable or recyclable
    o Also consider how packaging may better integrate and enhance distribution

# Waste as resource

Another strategy of circularity is reducing product and packaging waste further by converting it into a resource to be used by something else.

- Convert bio-waste into compost or biochemicals and—for example, unusable waste, such as food scraps and sewerage sludge, can be returned to nature to decompose and regenerate natural resources, or treated to produce energy in the form of biogas
- Convert technological waste for use outside the lifecycle—for example, ByFusion converts plastics that are difficult to recycle into concrete blocks for construction purposes[162]

**Examples**

- **Plastic waste into furniture**—One small company delivering a regenerative service is **Plastic Whale** in Amsterdam, who take tourists on "plastic fishing" tours through the city's canals, then use the collected plastics to make office furniture[163]
- **Biodigester agricultural waste-to-fuel**—Sistema.bio installs Biodigester machines in farms in Kenya, Mexico, India, and elsewhere that turn agriculture and livestock waste into fuel for homes. Waste is broken down inside the prefabricated modular biodigesters by bacteria, returning waste to its purest form which can then be used as fertiliser. The only other by-

product is methane which is run via tubing to the homes to be used as energy for cooking, lighting, running machines, etc.

# Design for post-use loops

**Design to keep materials in loops of ruse, repair, remake, and recycle.**
Using your *Product Anatomy Map* and *Lifecycle Map* as references, explore the following strategies to identify problems and brainstorm solutions.

## Collection

A key strategy for continuing the flow of materials from product use into circular loops is the *collection* of products from users, which must form part of the business model. Without easy and accessible collection processes, users are likely to discard their products with other waste headed for landfill or succumb to 'disposal sense of guilt' and keep unused products in forgotten drawers.

There are several methods for collection—including emerging ideas to combat 'disposal guilt'[164].

- Takeback programs
- Deposits and returns
- Gamification of returns
- Trade-in for credit

## Reuse/Repurpose

Reuse, repurpose, reimagine.

- Enable reuse by enabling ways to get them directly back to other users—like creating an online space to connect sellers and buyers
- Repurpose unwanted products before they require recycling—like adding Bluetooth to vintage radios
- Reimagine disposable products as reusable ones, like coffee cups
- Explore how any biological materials can be repurposed by the user, like using natural fabrics for rags or cushion stuffing
- Repurpose technological waste as material for future products

## Disassembly

To enable better repairability, refurbishment, and recycling, products need to be designed for easy and fast disassembly/reassembly.

Create a disassembly map to record:

- Order parts are removed
- Materials of parts
- Fasteners and adhesives used
- Tools needed
- Ease or difficulty of manipulation
- Any damage caused during disassembly
- Where the key components are in the disassembly process
- Time it takes to disassemble and reassemble

This is to assess for points of circular interventions, such as improvements to repairability, choosing more sustainable, safe, and ethical materials, and reducing waste.

- Preparation
    - To map disassembly, you'll need[165]:
        - Room to layout all parts
        - Tools (screwdrivers, pliers, wire cutters, etc.)
        - Measuring tape or ruler
        - Note capturing materials (pen, sticky notes, camera, an online whiteboard like Miro, Mural, etc.)
        - Depending on the complexity of the product, you might need containers of some sort to collect fasteners, etc. for each step (bowls, drinking glasses, etc.)
        - Depending on the size of the fasteners, etc., you might also need tweezers and a magnifying tool
        - Protection cloth, etc. for delicate components
        - Refer to manufacturer resources for guidance
        - Refer to *ifixit.com* for help and developing knowledge

- Disassemble
    - As you disassemble the product and layout all the parts, capture[166]:
        - A diagram flow of the parts and fasteners removed in order
        - Note the basic material type of each part, like wood, metal, plastic, coating, pigment, stone, engineered stone, foam, cotton, wool, nylon, etc.
        - Tools required at each step
    - Note disassembly difficulties:
        - Ease of handling
        - Clarity—clarity of what to do, visibility/identifiability/accessibility of connectors

- Non-reusable connectors (adhesives, causes damage when disassembled, etc.)
- Uncommon tools required

o Note key components regarding:

- Failure—highest failure rate or functional importance
- Environmental—the most environmentally harmful
- Economic—the components with the highest embedded economic value
- Recovery/reuse— high potential for reuse/recovery[167]

o Create a final list of all components in order of removal and their function
o Note the total time it took to disassemble

- Use the insight to discover opportunities for innovation:

  o Design for standard tools to be used that are accessible to both users and service providers
  o Optimise the accessibility of main target components for repair and replacement, cluster together components with similar life expectancy, and reduce steps required to get to them[168]
  o Ensure disassembly is non-destructive by avoiding adhesives and permanent fixings (welding, etc.) and using alternatives such as:

    - Click/snap solutions
    - Metal screws
    - Click fingers
    - Press fit
    - Shrink foil
    - Self-screwed/tapering or connectors
    - New adhesives that are "de-bonded" using light, heat, and magnetic field
    - Detachable components (for waste and pollutants)

  o Ensure materials are marked for easy identification (avoid labels or paint that may contaminate recycling)
  o Design ways to safely remove any hazardous materials that can't be replaced with safe alternatives
  o Include manufacturers, repairers, and recyclers in design workshops for their practical insight

For a deep dive into disassembly, see Philip's *Circular design for disassembly and repair*:

*engineeringsolutions.philips.com/webinar-making-circular-innovation-work*

### Refurbishment/Remanufacturing

Refurbished products return to the service provider to be renovated cosmetically and/or mechanically for reuse.

Remanufactured products return to the manufacturing process where some parts might be reused while others are recycled and remanufactured into new usable goods.

Include manufacturers in the design phase for practical insight and explore using the *Modulatory* and *Disassembly* strategies to make refurbishment and remanufacture easier.

### Recycle

Recycled products go back to the materials processor where the materials are sorted and prepared for reuse in new products.

To optimise for the recycle process:

- Use Sustainable and ethical materials
- Design in ways to keep materials uncontaminated
- Consider the recycling systems that exist in the locations where the product can be discarded and design post-use product *Collection* into the business model
- Design for *Disassembly*
- Include recyclers in the design phase for practical insight

# Sustainable and ethical digital

Incorporate data collection and digital platforms into your product/service at any lifecycle stage to enable[169]:

- Information sharing between systems and people for continuous feedback, agile iteration to help maintain circular integrity.
- Measuring and monitoring of circular changes and data, behaviours and attitudes, to continually optimise experience, circularity, and inclusivity
- Data-based behavioural design strategies, like sharing back the personal and/or collective energy saved, complimented with gamification
- Monitoring lifecycle resource and energy use

Digitising the product and service also supports the decentralisation of distributed design by allowing designs to be shared and goods to be made locally.

**Prompts:**

- What information do the actors of each stage have that they could share (data, insights. etc.) and with whom?

- Collect legal data about user behaviour, resource and energy use, etc. using sensors, surveys, data analytics, user-reported information, heat maps, opted-in tracking, or AI
- Use data to monitor a circular design change or constantly optimise—create a hypothesis of what you hope to learn, articulate what evidence you think you need, and plan how you will monitor that data
- Create a community hub using social media, or other online peer-to-peer tools to enable feedback between customers, between value web partners, or between customers and value web partners (also known as Bridging[170])

### Sustainable and ethical data:

- Embedding feedback must be done responsibly by managing issues around data protection and abuse
- Give users transparency and control over their data use
- Use *Sustainable and ethical digital* practices to reduce the carbon footprint of data use:
    o Optimise web content
    o Design for mobile first
    o Maximise user journey efficiency
    o Minimise emails and communications
    o Use the Foster Stewardship strategy to encourage sustainable behaviour
    o Utilise green hosting

# Foster user stewardship

### Enable life-centred user behaviour.

Taking cue from Indigenous custodianship of land, and utilising transparency and behavioural design, designers can emotionally connect users to the resources they use by informing and enabling them to:

- Maintain and repair the product whilst in their care
- Know how to properly move their products into circular loops via convenient end-of-use collection strategies
- Utilise traceability to be aware of the location and people that provided the product's resources, and to help establish empathy and care beyond the product's immediate use and benefits

Sustainable user behaviour can be fostered via behavioural design strategies.

### Conduct a Behavioural Diagnosis

- Decide on a behaviour you want to install or change—this should be a specific behaviour and measurable
- Define the relative journey by determining the start and end points—these should be wide enough to capture any influences occurring before key decision making
- Map the journey, detailing each behavioural step, each decision, obvious influences, and the jobs to be done
- If you're mapping with a user, note what they do and what they say as these won't always match, and noting both will help in understanding their personal biases in interpretation
- Layer with stakeholder insight and any user behaviour data
- As an option, you could also draw on service design and map the services and people supporting the customer journey along the way to identify any connections to undesired user behaviour

**Define barriers**

- Note the friction points that block the desired behaviour. These can be structural (environmental/functional aspects, such as low discoverability, small font, inadequate equipment, etc. that cause things like decision fatigue) or psychological barriers (decision paralysis, low brand trust, etc.).
- Four common barriers are:

  o Attention bias
  o Cognitive overload—too many options, too much thinking required to decide, etc.
  o Status quo—avoiding behaviour change due to familiarity with existing behaviour and/or a sense of loss when changing
  o Mental models—preconceived ideas of how things are or should be that drive existing behaviour or prevent change

- Refer to a great list of cognitive biases like *thedecisionlab.com/biases*
- Consider the need for personas or archetypes (such as "Conformists Vs Rebels" as defined by consumer insight specialists Kantar and their Millennium Monitor—watch from 9:52 at *youtube.com/watch?v=rbVDiU-bpGI*)

**Design interventions**

- Barriers are potential intervention points. Note and brainstorm any opportunities to increase motivation and change behaviours. Think in terms of:

- Remove or reduce barriers—Remove any barriers and friction points you can, and ease others by simplifying the decision-making processes
- Amplify personal, environmental, and social benefits to motivate users to complete the desired behaviour

  o Think in terms of functional, social, and emotional benefits
  o Ensure users know they are making a difference
  o Align benefits with values as much as any financial or convenience gain
  o For longer-term benefits (incentives), also create a perceived immediate benefit (reward)
  o Enlighten—use information, feedback, or means of reflection to influence a user's knowledge, values, attitudes, and norms
  o Spur—motivate a user towards a desired behaviour through expressing benefits unrelated to environmental outcomes
  o Use clear, meaningful, and jargon-free content
  o Offer meaningful and/or personalised recommendations

- Enable simplified action with sustainable choice as default

  o Clear navigation and non-distracting interface design
  o Step out information and actions and employ journey cadence
  o Provide assistance and make it easy to access
  o Provide rails of guidance
  o Trigger action at times of high motivation
  o Reduce time, effort, and cognitive load
  o Force—compel sustainable behaviour through limiting functionality or reducing the appeal of unsustainable or unethical behaviour
  o Match—adapt the behavioural design pattern to match the original and expected interaction

- Reward by providing:

  o Feedback on impact of sustainable aspect of user's action
  o Personal rewards to recognise the user's self-accomplishment
  o Social rewards, that can be shared with the user's tribes and networks
  o Variation of rewards to create excitement

- For actions/journeys that need to be repeated by users over time, turn the sustainable behaviours into habits by:

  o Acknowledging and celebrating progress
  o Learn from the user's ongoing behaviours to personalise the relationship between them, the product and business

- o Focus on leveraging intrinsic motivations that come from within self and foster long engagement by enhancing a user's:
- o Sense of purpose and accomplishment
- o Connection
- o Self-expression
- o Autonomy
- o Mastery

- Consider using gamification, but keep it meaningful and avoid leveraging too many extrinsic motivations that come from externalities, like money and status (it's okay to use these in small and meaningful doses)
- Create ways for users to invest and add value to the product/service to build the emotional connection
- Consider focusing on one pathway of sustainable behaviour that would best suit your users[171]:

  - o User's choice of what they use/purchase—choosing a primary artefact that consumes fewer resources during use
  - o Changing their ways of use:

    - Use the artefact in a way that consumes fewer resources or produces fewer pollutants
    - Users adapt their behaviour according to situations
    - User curtails the use that uses too many resources (energy, data, network usage, etc.)

  - o User maintenance and repair—encourage and enable both
  - o Embed mechanisms to help regulate use

- Consider how digitising data and/or making it intelligent can help with any of the above

**Test and iterate**

- Test concepts with a diverse range of peoples to determine which variation has the highest potential to positively impact attitudes and behaviours. Test the concepts within their full journey to check for other unintentional triggers or triggering the opposite effect to the desired one.

# 3.3
# Lenses

Whether designing the product or improving the supply chain, there are common strategies that can be applied as lenses at multiple lifecycle stages, and from micro to macro levels, to ensure design and business are constantly seeking opportunities to be more life-centred.

- **Nature**—Be nature inspired for sustainable design solutions
- **People**—Design for pluri-clusivity as default and for innovation
- **Animal**—Consider animals as legitimate stakeholders
- **Regeneration**—Take a systems view of your product, business, and supply chain to identify resources and communities to renourish
- **Energy**—Power everything with efficient and renewable energies
- **Time**—Scan alternate pasts & consider alternate futures to protect all past wisdoms and future stakeholders

Lenses could be used at any stage of design, including post-launch monitoring.

Refer to your *Lifecycle Map, Phygital Map*, or a specific design problem, use the lenses to assess problems from different perspectives.

Identify opportunities for interventions at different levels and consider how they may impact each other level.

Alternate the *Lenses* often to ensure life-centred thinking is maintained, whether zoomed in or out.

The following tools are offered as a guide to implementing the *Lenses*, but feel free to try others and experiment with how often and how deeply you implement them.

# Nature Lens

Consider nature as a legitimate stakeholder by viewing your work from nature's perspective.

## Identify environment stakeholders

Use the *Lifecycle Map* and/or the *Life-centred purpose* tool.

## Environment persona

**Environment personas are visual characterisations of the environments impacted by a product's lifecycle.**

*Tool—Environment Persona*

*Download from lifecentred.design*

After you've identified environment stakeholders, create a persona for each one to enable empathy and consideration for them during design.

Anthropologist, UX researcher, and environment-centred design advocate, Monika Snezl advises that environment personas differ from user personas in that they are primarily based on facts, so be sure to check and double-check the quality of facts included[172].Data and statistics for environment personas can be sourced from documentation from respectable global organisations like the UN, from more localised affiliate-free organisations, and experts.

## The Environment Persona

The persona (Figure 24 - Environment persona) consists of three main sections:

- **Image and quote**—This section provides a summary for instant empathy. Add an image and a quote 'by the environment' that captures both their individuality and significance to the ecosystem. Avoid over-humanising, use terms that reflect their values and world.
- **Lifecycle impacts**—Tick the boxes representing the product lifecycle stages that potentially impact the environment.
- **Needs, challenges, solutions**—Summarise their needs and the challenges to their thriving existence. Keep these related to the lifecycle stage where they are impacted. Research Prevention and Healing solutions to the challenges—this makes the persona a practical tool for brainstorming about improvements or regeneration.

Refer to the persona as needed during the design process and keep it visible as a constant reminder to consider them in all decision making.

# DAMIEN LUTZ

## Trees

> We provide oxygen and limit carbon in the atmosphere. We reduce air pollution, provide food and shelter for wildlife, minimise erosion and maintain healthy soil, increase rainfall, and absorb sunlight as energy.
>
> 80 percent of all terrestrial plants, insects, and animals call forests home. Nearly one third of people in the world depend directly on forests for their livelihoods.
>
> We communicate and collaborate together, and we share our fear about our space and soil for thriving becoming too unfit.

◉ Material extraction | ☐ Supply chain | ☐ Product in use | ☐ 2nd life | ☐ Waste

**NEEDS**
- Reduced deforestation
- ~~Time to grow~~
- Healthy soil

**CHALLENGES**
- Deforestation
- Urbanisation
- ~~Climate change~~
- ~~Increased wildfires~~
- Invasive species
- Habitat degradation

**PROTECTION**

| PREVENTION | HEALING |
|---|---|
| Use less paper | Plant a tree |
| Recycle paper and cardboard | Practice eco-forestry |
| Use recycled products | Raise awareness |
| Buy only sustainable wood products | Respect the rights of indigenous people |
| ~~Don't buy products containing palm oil~~ | ~~Support organisations fighting deforestation~~ |
| ~~Reduce meat consumption~~ | |
| Respect the rights of indigenous people | |

Non-Human Environment Persona. Damien Lutz 2022. www.futuringdesign.com auted. NonCommercial 4.0 International (CC BY-NC 4.0)

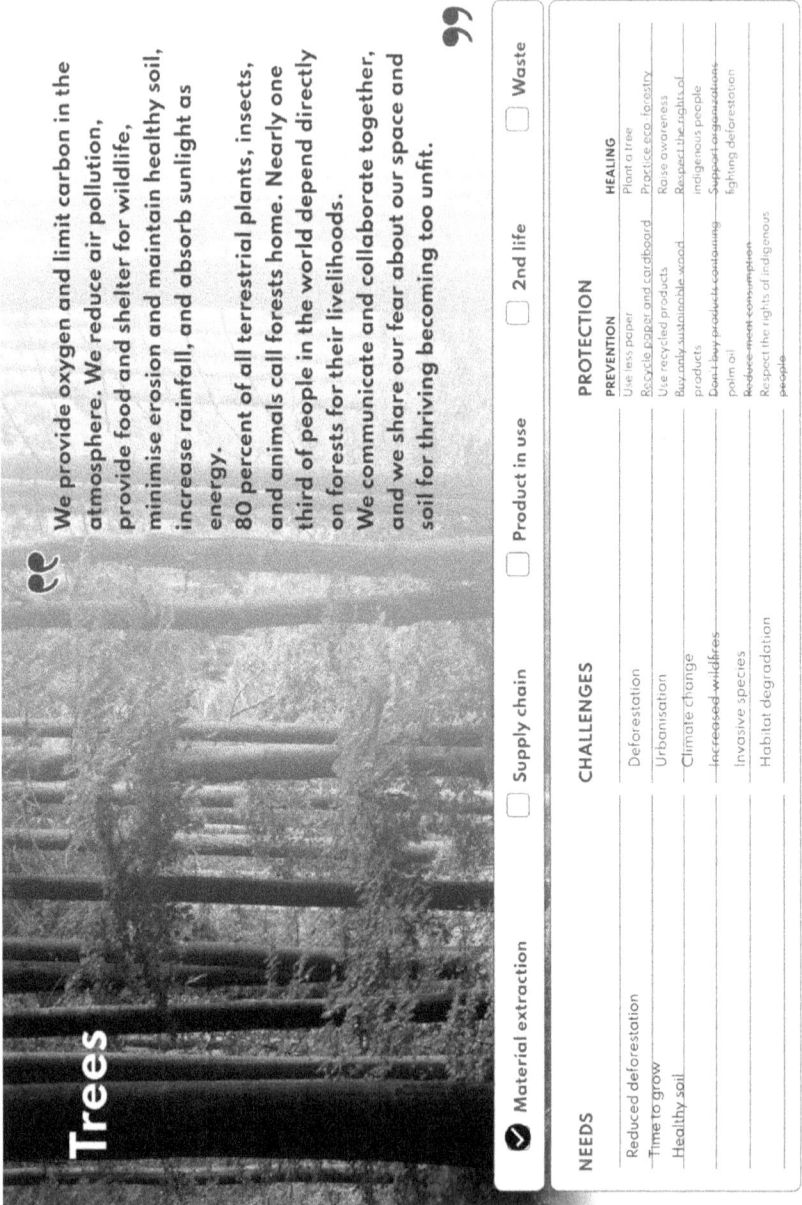

Figure 24 - Environment persona

214

## Use biomimicry for sustainable design solutions

Tool–Bio-inspired solution tool (Figure 25 - Bio-inspired solution)

*Download from lifecentred.design*

The natural forms, functions, and systems in nature are the result of billions of years of evolution—we can draw from nature's solutions to inspire efficient, sustainable, and regenerative designs by implementing biomimicry.

The scales of nature's solutions range from molecular, cellular, organismal, to community and landscape, but all can inspire solutions for digital and physical products.

Steps to activate the *Nature Lens*:

- **Define the problem** as basic functions of the required solution:

    o  In the example, the problem is 'My product can't get wet'. To reduce the problem to a basic function, consider the function which the solution would need to perform. Keep it simple, like *building, moving, heating, cooling*, etc,—it should read like it could relate to anything human, machine, or of nature. In this example, the function is '*repel water*'.

- **Search** a resource like '*asknature.org*' or 'ultimate-guide-to-genius-of-place' for biological organisms and systems that perform the same functions
- **Explore** how the organism or system achieve the solution with its form, functions, and/or by being part of a larger ecosystem
- **Interpret** the biological solution as design principles
- Using the design principles as a guide, sketch out an **engineering solution**
- **Optimise** the solution for circularity and regeneration by using Biomimicry 3.8's Life Principle's as a checklist—*biomimicry.net/the-buzz/resources/designlens-lifes-principles*

## Bio-inspired solution for:
Plastic parts for bluetooth headphones

**VALUE ADD**
Music, information, entertainment

**DESIGN PROBLEM**
My product can't get wet

**Solution function**

Repel water

**Nature's solution**

Cicada Wings Repel Water

When viewed under a microscope, arrays of rows can be seen of tiny pillars poking up from wing surface, which leave no for water to adapt to, so water runs off

**Design principles**

Surface to have irregularities that are smaller than water droplets

**Design solution**

Use a material that has surface irregularities that are smaller than water droplets

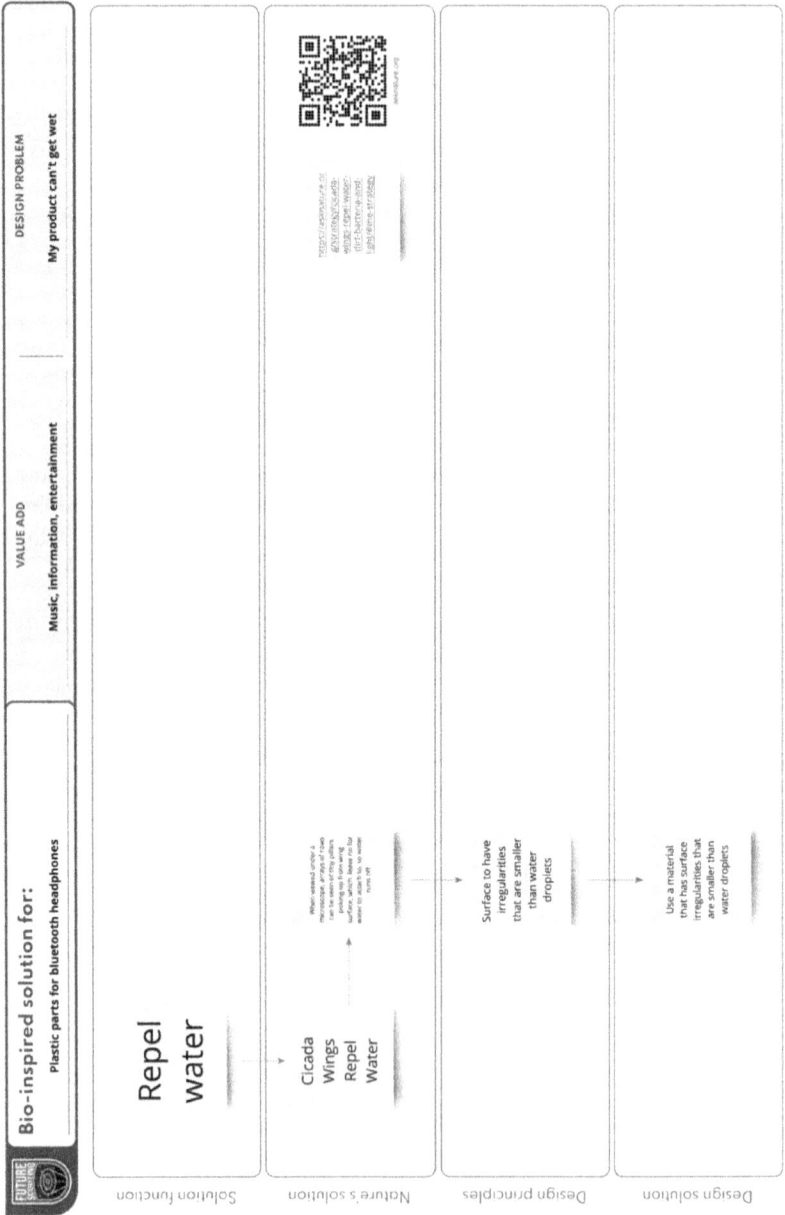

Figure 25 - Bio-inspired solution

# People Lens

**Design for pluri-clusivity as the default experience, and for innovation.**
Draw from the *Inclusive and Pluriversal Principles* to decolonise design processes, remain vigilante to the power of privilege, and foster the celebration of diversity, equality, and differing ways of knowing and being into the solution, business, partnerships, and processes... by default.

**Key considerations:**

- **Know** the context of other's experiences (users, partners, team, invisible humans—all peoples)
- **Know your power** and use it to help others
- **Gather** all the right people (as determined by a *Lifecycle Map*) and ensure they are included at key decision making and testing times[173]: *Set up, Diverge, Converge, Wrap up*
- **Allow modern and traditional knowledges to merge**
- **Let the people design**
- **Use the complexity** of pluri-clusivity as inspiration to innovate inclusive, adaptable, and joyous experiences

## Identify all human stakeholders

Use the *Lifecycle Map* and/or *Life-centred Purpose* tool to identify all peoples impacted by the lifecycle, then use these tools to activate the *People Lens*:

- **Non-user Persona**—Create personas to represent the invisible humans impacted by the lifecycle
- **Power Pixel (from Exercise 2)**—A tool for designers and stakeholders to explore how to use their privileges, values, biases, and assumptions as powers for positive change
- **Life-centred Workshop**—Infuse human-centred design workshops and design jams with pluri-clusive principles to decentre the designer and allow participants to shape the process through participatory and inclusive co-design

# Poonam
## Child worker

" I am 12 years old, some of my friends are younger. We can not go to school, we have to make money for home. I am so tired, I squat down all day, and the fumes make me sick. But if I don't go, I feel guilty, because my parents can't get work. I love my parents, they look after me. Some of my fiends are not so lucky. I used to think, one day, I will make a lot of money, and I will go to school, but I don't think so now. "

❯ Material extraction ❯ Supply chain ☐ Product in use ☐ 2nd life ☐ Waste

**NEEDS**

Food, clean water, shelter, healthcare
Protection
Development
Community participation
Time to play

**CHALLENGES**

Poor living conditions
Low levels of income
Lack of job diversity

**PROTECTION**

| PREVENTION | HEALING |
| --- | --- |
| Check material sources | Raise awareness |
| Do not source from areas known | Sponsor a child |
| for child labour | Make a donation |
| | Connect with humanitarian programs |

Figure 26 - Non-user Persona

## Non-user Persona

*Tool: Non-User Persona (Figure 26 - Non-user Persona)*
*Download from lifecentred.design*

**Instructions:**

The Non-user Persona consists of three main sections:

- **Image and quote**—This section provides a summary for instant understanding and empathy. Add a representational image of the non-user stakeholder and a quote, written in first-person by the non-user that captures both their individuality and how they are impacted. Use terms that reflect their values and world.
- **Lifecycle impacts**—Tick the boxes representing the product lifecycle stages that potentially impact the non-user.
- **Needs, challenges, solutions**—Summarise their needs and the challenges to their thriving existence. Keep these related to the lifecycle stage where they are impacted and localised if possible. Research solutions to the challenges and split them into Prevention and Healing—this makes the persona a practical tool for any brainstorming sessions about improvements or regeneration

## Power Pixel

*Tool–Powel Pixel (Figure 27 - Power Pixel)*
*Download from lifecentred.design*

**Know the power of your privileges, values, assumptions, and biases.**

As pluriversal design shows us, designers need to be aware of their power and the potential impacts their work can have by understanding the power of design, how it's used, and by whom—from deciding who and what knowledge beliefs are included or excluded in research, design, and decision making, to how design influences the sourcing and production of the materials that impact different places and peoples.

The *Power Pixel* is a kind of mindfulness token for designers and individuals to hold in their hand to keep the power of design high in their consciousness and to remember their place in context.

**About the pixel**

There are four sides of the Pixel to complete:

- I will lend and call out PRIVILEGES
- I will look through EXPERIENCE LENSES
- I will champion MY VALUES
- I will challenge MY ASSUMPTIONS

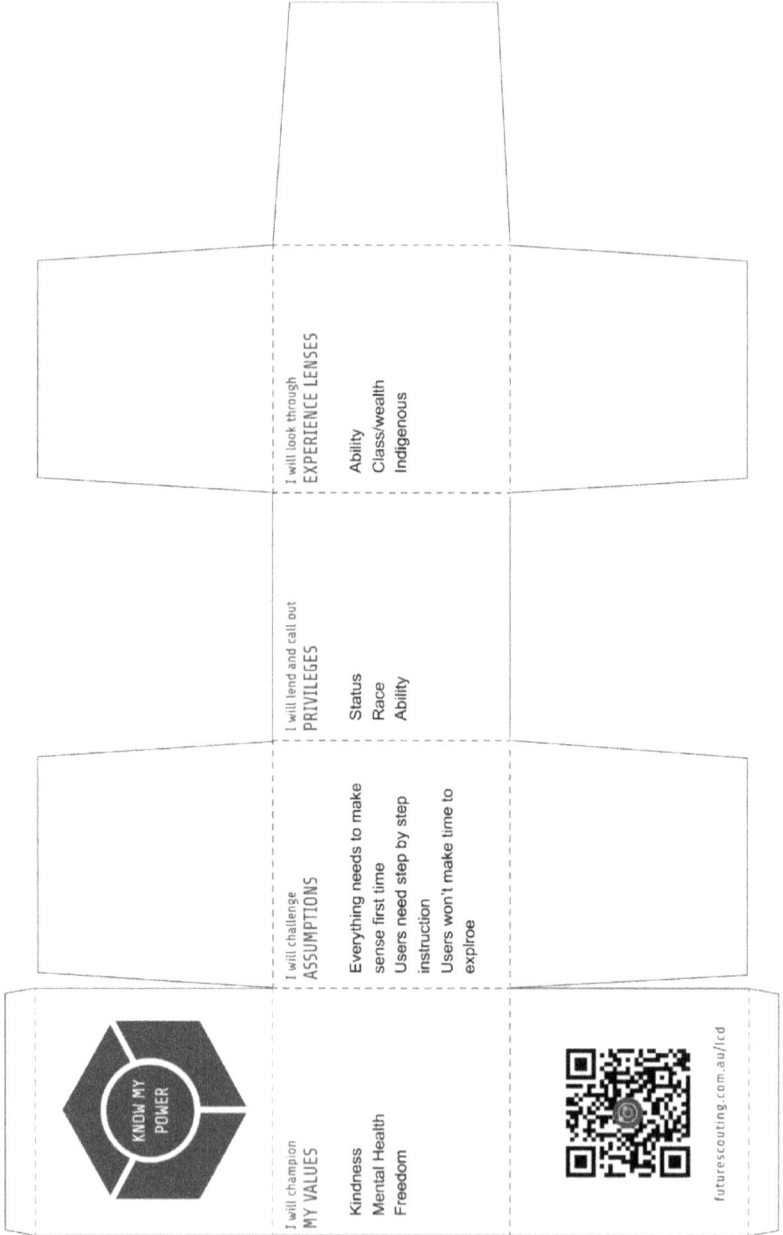

Figure 27 - Power Pixel

**Determine your Values**

Core values are our highest priorities and our deepest beliefs. Once you know your values, you can use them to guide your behaviours, decisions, and actions to ensure you are creating solutions that will foster the future you want

- Think of three good unforgettable experiences, choose value words to describe them
- Think of three unforgettably bad experiences, choose value words to describe them, then flip them to their opposite
- Think of three people you admire, choose core value words to describe them
- Think of three people who really challenge your values, choose value words to describe them, then flip them to their opposite
- Think about the things in your life and name the top three you couldn't live without. Associate the feeling of them with a core value.
- Group all the similar values to identify the three most important to you.

Other methods to determine your core values:

- Values discovery activities by CEO Sage— *https://scottjeffrey.com/personal-core-values/*
- Core value worksheets by Tommi LLama— *https://tomillama.com/personal-core-values/*

**Determine your Privileges**

Privileges are the unearned, identity-based benefits we inherit just by being part of one group, that we can often be unaware of, and are enjoyed at the expense of others. Knowing them allows us to use them to help the less privileged, or to denounce them to rebalance inequity

The Privilege Web* in the *Power Pixel* template guide captures key privileges and allows you to mark yourself against each.

- Map the size of your privilege:

    o  Go around the web and mark where your privilege is for each segment
    o  Use the blank segments to add any other privileges you think should be included (if you leave them blank, mark yourself at the highest level/outer ring for these)
    o  Draw lines between each mark on the privilege segments, and colour in the emerging shape with a highlighter (as per the example) to reveal the size of your privilege

- Understand the impacts of your privileges

    o  In the first outer ring, 'PRIVILEGE BENEFITS', use green sticky notes to capture how your privileges benefit you

- In the inner circle, 'OPRESSION EFFECTS', use red sticky notes to capture how your lack of privilege affects you, and how your privileges affect others

- Think of ways to mitigate the negative effects

  o In the outer-mist circle, 'MY POWER', brainstorm how to use your benefits as powers to:

    ▪ Help those less privileged
    ▪ Help the more privileged see what they have to lend

- Consider which segments most relate to your values
- Choose a few ideas and write these on the **I will lend and call out PRIVILEGES** side of the Pixel

You could also add the people most affected by your privileges to your **EXPERIENCE LENSES**.

*The Privilege Web was based on the great power literacy work of Maya Goodwill in her *Field Guide to Power Literacy*. I highly recommend using the *Field Guide's* exercises to build up 'your awareness of, sensitivity to, and understanding of the impact of power and systemic oppression in participatory processes[174].'

**Determine your Experience Lenses**

The Experience Lenses represent the human experiences unlike your own, to see the design experience through their eyes, to lend your privilege to as a positive power, and to ultimately design solutions that celebrate the full spectrum of human experience and ways of being.

- Consider how your product might impact people who:

  o Experience life differently from you
  o Are often not included, misrepresented, or marginalised by mass design
  o You may have bias against
  o You just want to have more empathy for

- Think in terms of:

  o Accessibility
  o Gender identity
  o Sexual orientation
  o Culture
  o Age
  o Body type
  o Personal belief
  o Pluriversal experiences

- Consider which segments most relate to your values—for example, if one of your core values is mental health, you could include the consideration of users on different mental health spectrums
- Consider which segments most relate to the SDGs identified in your *Life-centred Purpose* tool

### Determine your Assumptions

We all have established opinions that generate automatic responses affecting our thinking and behaviours. These assumptions can negatively influence designer's choices of solutions and participants.

For the *Pixel's* purpose, understand your personal assumptions specific to the theme of the project's purpose so that you can challenge the assumptions you default to.

- Consider:
  - When you think of the theme, what do you automatically assume about how the solution might need to be—style, materials, order of experience phases, etc.?
  - What do assume users and stakeholders think and need?
  - What do you assume can't be done?

Group your assumptions into a few key insights and write these on the assumptions side of the Pixel.

How to use the Pixel

Once completed, print out your Pixel, fold it into its cubic shape (no adhesive required), keep it on your desk during the project, and define your ritual of use:

- At the start of each day
- Before any meetings, design sessions, or research sessions
- At times of decision making, confusion, or frustration
- Any time you feel you need to tune into your place in context

By its tangible nature, the *Pixel* prompts designers to pause and tune into the power of their privileges, values, assumptions, and biases as they design to help create kinder and more mindful experiences for all humans, animals, and planet.

Our core values remain the same over longer periods, but biases and assumptions can change with each project, so you might want to create new *Pixels* over time or per project.

## Life-centred workshop

**Infuse workshops with pluri-clusivity by default and use diversity's complexity to inspire innovation.**

Tool–Life-centred workshop cheat sheet

*Download from lifecentred.design*

Infuse design jams and workshops with pluri-clusive principles to allow participants to shape the process through participatory co-design.

**Workshop A—Solving a location-specific problem**

Below are tips for a life-centred workshop for a location specific design problem, or for workshopping with the supply chain partners based in a different location to where the business is based.

**1. Prepare**

- Consider the accessibility of your workshop's location and any physical, mental, cultural, or identity barriers—use a mix of communication methods, and do your best to ensure tools, images, terminology, etc. used in the workshop are culturally appropriate and allow all to fairly participate
- Be mindful of the dominant narratives (colonial, heterosexual, cis, fully abled, capitalistic, hegemonic) to better see and understand alternate values
- Be mindful of one's own values, place in context, and any privileges, and use these to aid others
- If you're designing for a particular product design problem (repairability, durability, etc.) refer to the relevant strategy for task ideas

**2. Connect to locality**

- Know context/place

  - Research the location, its history, people, and resources
  - Matchbox Studio recommend using the 'art of noticing[175]':

    - Walk around the location, taking photos and notes of what you see, smell, hear, feel
    - Analyse these to discover patterns, learn about local values and biases

- Systems map for margins of impacted people
- Recognise past knowledge, wisdom, struggles, and ways of existing as a human

  - Recognising past knowledge and wisdom allows it to be retained to form 'at least a hybrid of modern and traditional knowledge'
  - Revisit the past of a problem to discover ongoing drivers to consider when speculating about the future

When you have connected to locality, you'll be more empowered to...

### 3. Gather the right people

- The marginalised and/or misrepresented
- People from across regions and from diverse human experiences
- Indigenous Peoples to inform with traditional knowledge and experience
- Someone who knows the local resources
- Stakeholders
- Experts

### 4. Facilitation

Clarify roles:

- As the designer, you'll use you design knowledge to **facilitate** co-designing with the participants' values
- Technical experts, local knowledge experts, and traditional knowledge keepers will **mentor and guide the co-creation** with a mix of traditional and modern knowledge
- Participants will design, interact, talk, and share stories

Connect participants to place:

- Start with local representatives immersing the group in local lore, history, and awareness of local commons

Co-design:

- Keep the framework loose and allow it to respond to the participants
- Adapt and evolve according to the uniqueness of the people and project by having various viscerally engaging methods, action, and expression to inspire creative thinking:
  - o Poetry, music, sand drawing, sand structure building, writing, storytelling, world-building, rapid prototyping, animations, video,
  - o Dice, cards, Lego, string, etc.
  - o Gamification
  - o Encourage discussion, interaction, and storytelling (also known for healing)
  - o Create visuals of desires/archetypes/ to use as guidance
- Let the commons and history influence the design and bring the community together
- Allow multiple threads and narratives to emerge rather than trying to converge on one
- Encourage exploration of pleasure and joy as much as needs
- Use inclusive content (images, terminology, etc) that are culturally appropriate and respectful of the diversity

Experiment with strategies & lenses:

- *Localise* with local resources and makers
- Explore *Nature* for form, function, and system solutions through biomimicry
- Consider the impacts of your solution, both good and bad, over *Time* through forecasting
- Explore how to *Foster user stewardship* using behavioural design

**5. Playback, Critique, and Voting:**

- Allow everyone time to present their ideas
- Allow everyone to have a voice in sharing their thoughts
- Allow everyone to vote on the ideas they think are the best

**Workshop B—Solving a digital problem**

This is very similar to Workshop A, but without the location-specific components.

**1. Prepare**

- Consider the accessibility of your workshop's location
- Be mindful of the dominant narratives (colonial, heterosexual, cis, fully abled, capitalistic, hegemonic) to better see and understand alternate values
- Be mindful of one's own values, place in context, and any privileges, and use these to aid others

**2. Gather the right people**

- The marginalised and/or misrepresented
- People from across regions and from diverse human experiences
- Indigenous Peoples to inform with traditional knowledge and experience
- Stakeholders
- Experts

## 3. Facilitation

- Clarify roles:

  - As the designer, you'll use you design knowledge to facilitate co-designing with the participants' values
  - Technical experts and traditional knowledge keepers will mentor and guide the co-creation with a mix of traditional and modern knowledge
  - Participants will design, interact, talk, and share stories

- Co-design:

  - Keep the framework loose and allow it to respond to the participants
  - Adapt and evolve according to the uniqueness of the people and project by having various viscerally engaging methods, action, and expression to inspire creative thinking:

    - Poetry, music, sand drawing, sand structure building, writing, storytelling, world-building, rapid prototyping, animations, video,
    - Dice, cards, Lego, string, etc.
    - Gamification
    - Encourage discussion, interaction, and storytelling (also known for healing)
    - Create visuals of desires/archetypes/ to use as guidance

  - Allow multiple threads and narratives to emerge rather than trying to converge on one
  - Encourage exploration of pleasure and joy as much as needs
  - Use inclusive content (images, terminology, etc) that are culturally appropriate and respectful of the diversity
  - Experiment with strategies & lenses
  - Explore *Nature* for solutions to digital system structures
  - Consider the impacts of your solution, both good and bad, over *Time* through forecasting
  - Explore how to *Foster user stewardship* using behavioural design

## 4. Playback, Critique, and Voting

- Allow everyone time to present their ideas
- Allow everyone to have a voice in sharing their thoughts
- Allow everyone to vote on the ideas they think are the best

# Animal Lens

**Consider animals as legitimate stakeholders by viewing your work from an animal's perspective.**

When factoring in the welfare of animals in your designs, consider whether you are protecting their existence, designing something they could interact, or both.

- Protect—Designing solutions for human problems that protect animal ways of living
- Engage—Design solutions to be used by animals, whether via human-led or animal-initiated experiences, and for specie-to-specie and cross-species interaction

### Identify animal stakeholders

Use the *Lifecycle Map* and/or the *Life-centred purpose* tool.

### Observe and empathise

- If you have a pet or legal access to safe interaction with an animal, practice empathy by allowing the animal to initiate and lead play, following its lead, and exploring how that interaction is different from the usual experience of human-led interaction.
- Observe to identify[176]:

  o Similarities and mismatches with human behaviours can then be used as prompts for innovation by exploring:
  o Trust factors between humans and animals
  o How animals complete the same or similar tasks to humans
  o Barriers to animal accessibility to engage/interact with humans and the human world
  o The animal's types of interaction, and which are most common (noise, licking, touching, brushing against others, etc.)
  o How animals sense and navigate environments, and how these abilities change with different environments
  o What elicits different types of responses

## Animal personas

*Tool: Animal Persona (Figure 28 - Animal Persona)*
*Download from lifecentred.design*

**Animal personas are visual characterisations of the animals impacted by a product's lifecycle.**

After you've identified animals to consider from your *Lifecycle Map*, create a persona for each one to enable empathy and consideration for them during design. I designed this template based on the aspects of animals that experts noted in their research so the persona can be used for protecting the animal's needs, habitat, and way of life, and as a reference for animal-computer interaction design.

## The Animal Persona

The Animal Persona (Figure 28 - Animal Persona) consists of three main sections:

- **Image and quote**—This section provides a summary for instant understanding and empathy. Add a representational image of the animal and a quote, written in first-person 'by the animal' that captures both their individuality and significance to the ecosystem. If the persona represents one individual animal, infuse the quote with their character and nature. Avoid humanising them too much by speaking with terms that reflect their values and world.
- **Protect**—This section informs designing products and experiences in a way that recognises and minimises potential impacts to animals. Summarise the habitat and environment they need to thrive, their needs and joys, and challenges to their thriving existence. Tick the boxes representing the product lifecycle stages that potentially impact the animal.
- **Engage**—This section informs designing experiences that the animal with engage or interact with. Summarise how they navigate, communicate, and interact with the world. Clarify their barriers to engagement and inclusion with the human world and identify any key humans they interact with.

Refer to the persona as needed during the design process and keep it visible as a constant reminder to consider them in all decision making.

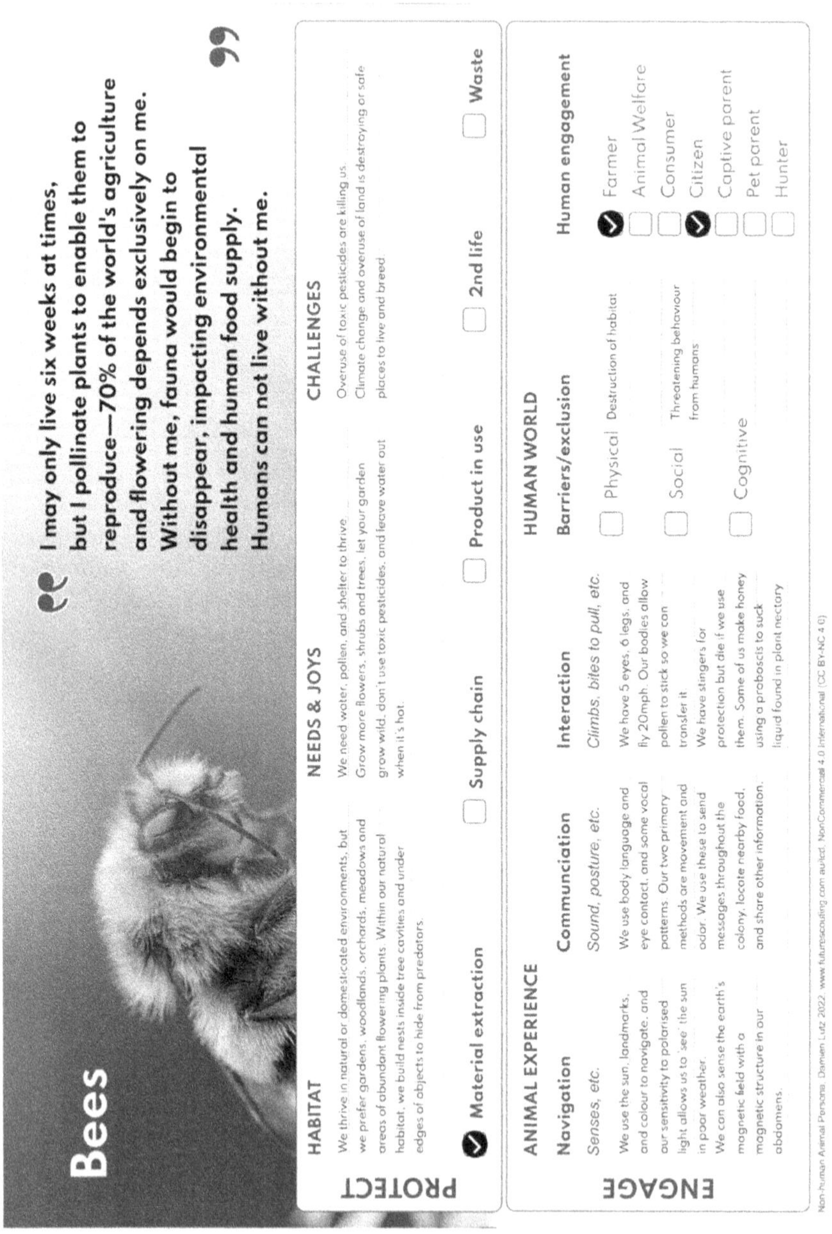

## Bees

> I may only live six weeks at times, but I pollinate plants to enable them to reproduce—70% of the world's agriculture and flowering depends exclusively on me. Without me, fauna would begin to disappear, impacting environmental health and human food supply. Humans can not live without me.

### PROTECT

**HABITAT**
We thrive in natural or domesticated environments, but we prefer gardens, woodlands, orchards, meadows and areas of abundant flowering plants. Within our natural habitat, we build nests inside tree cavities and under edges of objects to hide from predators.

**NEEDS & JOYS**
We need water, pollen, and shelter to thrive. Grow more flowers, shrubs and trees, let your garden grow wild, don't use toxic pesticides, and leave water out when it's hot.

**CHALLENGES**
Overuse of toxic pesticides are killing us. Climate change and overuse of land is destroying or safe places to live and breed

☑ Material extraction  ☐ Supply chain  ☐ Product in use  ☐ 2nd life  ☐ Waste

### ENGAGE

**ANIMAL EXPERIENCE**

**Navigation**
*Senses, etc.*
We use the sun, landmarks, and colour to navigate, and our sensitivity to polarised light allows us to 'see' the sun in poor weather. We can also sense the earth's magnetic field with a magnetic structure in our abdomens.

**Communication**
*Sound, posture, etc.*
We use body language and eye contact, and some vocal patterns. Our two primary methods are movement and odor. We use these to send messages throughout the colony, locate nearby food, and share other information.

**Interaction**
*Climbs, bites to pull, etc.*
We have 5 eyes, 6 legs, and fly 20mph. Our bodies allow pollen to stick so we can transfer it. We have stingers for protection but die if we use them. Some of us make honey using a proboscis to suck liquid found in plant nectary

**HUMAN WORLD**

**Barriers/exclusion**
☐ Physical Destruction of habitat
☐ Social Threatening behaviour from humans
☐ Cognitive

**Human engagement**
◑ Farmer
☐ Animal Welfare
☐ Consumer
◑ Citizen
☐ Captive parent
☐ Pet parent
☐ Hunter

Figure 28 - Animal Persona

### *How Might We?* cards

*Tool: Interspecies How Might We? Cards*

*Access from lifecentred.design*

This deck of 16 cards provides prompts for interspecies design thinking to foster innovation and is split into two sections—PROTECT and ENGAGE. Use them when brainstorming design or regenerative ideas, such as when using the *Life Centred Purpose* tool.

# Regeneration Lens

**Take a systems view of product, business, and value web to identify resources, communities, people, and animals to renourish with every design, business, and value web decision.**

Zooming out to see the product, lifecycle, or a lifecycle stage as part of bigger systems, from a non-linear perspective, exposes previously unseen and under-appreciated connections that can be better utilised to enhance your life-centred purpose.

Detailed system mapping requires a learned understanding of its complexities. However, as an introduction to system thinking, you can use the elements revealed from your *Lifecycle Audit* and/or *Phygital Map* and use system mapping to explore non-linear connections by nurturing the people, animals, and environments of your product's ecosystem.

## Systems map for opportunities

System map your product, lifecycle, or a lifecycle stage to activate the *Regeneration Lens.*

### 1. Prepare

- Choose your problem—are you taking a holistic view of the entire lifecycle, or do you want to focus on one area that needs life-centred focus more than others?
- Gather the right people, including as many relevant users, stakeholders, value web actors, etc. as possible. You might also want to use animal personas to represent the environmental stakeholders, and future personas, as a reminder to consider future stakeholders
- Using human-centred design techniques, collaboratively define the problem, estimate a boundary, and reframe the problem as a question to investigate (not too broad, not too specific)

### 2. Map your system

- Define the elements/nodes—draw from *your Lifecycle Audit Map* (and any other research) and write on sticky notes the elements of your lifecycle system (representations that describe who, what where and when, these can be tangible and intangible, social, environmental, economic, and technological)

    o Consider social, technical, environmental, economic, and cultural elements, and infrastructure, policies, etc. that your lifecycle is connected to either by influence or dependence
    o If you're not sure if an element is essential part of the system or not, include it anyway and the process will determine that as you go
    o Arrange the notes in proximity to each other regarding their amount of interaction and importance to each other

- Map the relationships—Consider how elements change or influence each another in some way, like by moving resources (materials, energy, information, money, documents, etc) through the system?

    o Capture these relationships by drawing arrows from the component causing the effect to the one being affected. Use solid lines for a strong influence, and dotted lines for weaker ones. Label arrows with the resources being moved.
    o This is where it will begin to get messy and complex, continue with it for now
    o Look for non-obvious relationships by questioning the ones you didn't immediately connect
    o Note any feedback loops, where the flow of resource/influence leads back to its original causing element.
    o Fill in any information gaps you can

- Re-clarify the system purpose and boundary

    o What are the key components—which ones, if removed or blocked, would make the system falter, or weaken?
    o What flowing stocks (resources, materials, information, etc.) are important?
    o Remove any components or stocks that aren't essential to the system purpose

- Gather more insight—look at the system through:

    o Lenses
    o Your Life-centred Purpose
    o Use the *Strategies* as prompts, for example:

- - Are there chain effects of user behaviour that are impacting invisible humans or the environment in unexpected ways, and can this be changed with *Foster User Stewardship*?
  - *Optimising use*—can parts, resources, processes, etc. be reused or combined?
  - *Waste*—Look at each material and energy source, etc.—can any waste be reused as a resource by another part of the system?

  o Zoom in and tease out details, create new maps

- Define problems and leverage points as problem statements:

  o Look for areas of concentrated connectivity
  o Look for intervention points—think in terms of breaking bad cycles, strengthening relationships, adding buffers or reducing delays, adding rules, adding value
  o Note areas you can do something about—people, place, technology, UX, UI, etc.

- Use *How Might We*? statements to turn problems into questions for innovation
- Brainstorm ways to activate these leverage points by considering how the value web partners and your business value and/or life-centred purpose can heal damage, nurture potential, and innovate through:

  o Regenerating local jobs
  o Regenerating any damaged or ailing environment
  o Fostering diversity and inclusion
  o Healing injustices
  o Connecting locals with nature
  o Connecting locals with each other
  o Sharing knowledge and skills with the youth
  o Fostering preservation of local knowledge
  o Favouring the use of local resources over imported
  o Converting waste to energy/nourishment for something else

- Take our cues from nature with the Nature Lens to simulate the regenerative processes around us
- Narrow down, and keeping in mind what you can influence, and determine which life-centred *Strategies* will help enable these solutions.
- Use the *Gather and prioritise* tool to sort and prioritise your ideas
- Iterate solutions

  o Review the impact of your ideas by checking their effect on feedback loops

o Tweak and iterate to refine your intervention model
o Check future impacts with the *Time Lens*

- Iterate, launch, and monitor

# Energy Lens

**Power everything with truly sustainable energy sources.**

With the energy combined from all stages of the supply chain, the energy debt of a product lifecycle adds up.

Seek ways to power the entire supply chain with truly sustainable energy sources.

Renewables come from natural sources that are constantly replenished. With the planet being 71% water and constantly drenched in sunlight, sunlight and water are considered renewables. Wind, too, is renewable, although less reliable in some areas. However, wind turbines cause noise pollution and death and injuries to birds which in turn impacts ecosystems.

Truly sustainable means both the use and production of the source of energy must be sustainable and have no or minimal impact on peoples, animal, and planet.

Whatever source of energy you use, you can then seek ways to regenerate the resources used to produce and supply that energy.

For example, if your main source of energy is from solar power, you could invest in projects seeking to make solar powers more reusable, repairable, or recyclable.

Below are prompts to activate the *Energy Lens*:

- Check for any localised incentives and switch your business power to a supplier generating from renewable sources
- Look for any energy produced during the lifecycle that can be captured and reused
- Use *Sustainable and ethical digital* to assess your digital channels to reduce their energy use:

  o Optimise web content
  o Design for mobile first
  o Maximise user journey efficiency
  o Minimise emails and communications
  o Use the Foster Stewardship strategy to encourage sustainable behaviour
  o Convert to green hosting

Use your commitment to green power to:

- Support the global transition to renewables
- Improve your green business and product ratings

- Gain sustainable and environmental certifications
- Showcase your environmental commitment to your customers to foster their own sustainable choices
- Regenerate the resources required to produce your energy source

# Time Lens

**Use past perspectives to inform future considerations.**

Use foresight tools and methods to envision impacts of decisions, ideas, etc. on stakeholders while considering alternate pasts, to ground design with hindsight and empower it with informed speculation.

Below are some key foresight tools, followed by methods to apply the tools.

## Tools

- **Scope Wheel**—Determine the signals that are most relevant to your forecasting subject
- **Back glance**—Reflect on the past and how it may impact the future
- **Futures Wheel**—Brainstorm the future impacts of a decision, change, or idea
- **Futures Headlines**—An impactful way to summarise future speculations

## Methods

- **Future scenarios**—Mapping multiple possible future scenarios
- **Product resilience**—Exploring how a product may fair in multiple future scenarios
- **Strategy resilience**—Exploring how a strategy may fair in multiple future scenarios
- **Future stakeholders**—Explore future impacts on stakeholders to inform more life-centred design and forward-thinking regeneration strategies.

## A Foresight method

Explore the impacts of your future arcs to guide more life-centred design and decision making today:

### Product resilience

Explore the resilience of the current state of your product by testing how it fairs in the various future arc scenarios as created with *Future Headlines*.

- Identify the key need for your product to succeed

- Use the *Futures Wheel* to brainstorm how the headlines of each future arc impact the need:
  - Start with one future scenario headline
  - Write the key product need in the centre of the wheel
  - Place the headlines around the product
  - Imagine this occurs in the future and brainstorm and/or research how each headline impacts the need—positive, negative, or neutral—until you get a lot of ideas
  - Cluster all the ideas into themes, and write a summary of each theme to reveal key insights of how your product need might be impacted or change in each future scenario

- Repeat the steps above for each future arc
- Analyse all four future versions of the product need by looking for any commonalties across the futures that you might nurture or reduce

**Strategy resilience**

Explore the resilience of your life-centred strategy by testing how it fairs in various future arc scenarios.

- Identify the key need for your strategy to succeed
- Use the Futures Wheel to brainstorm how each future scenario impacts the need
  - Start with one future scenario
  - Write the key strategy need in the centre of the wheel
  - Place the headlines around the strategy
  - Imagine this occurs in the future and brainstorm (and/or research) how each headline would be impacting the need. These can be positive, negative, or neutral. Brainstorm consequences and combinations until you get a lot of ideas
  - When you have a lot, cluster them into themes, and write a summary of each theme to reveal key insights of how your strategy need might fair in that future scenario

- Repeat the steps above for each future arc scenario and analyse by comparing futures and looking for anything common across the futures that you might nurture or aim to reduce to strengthen your life-centred strategy

**Future stakeholders (people and planet)**

Review the people and planet stakeholders on your *Lifecycle Map* and choose one.

- Use the *Futures Wheel* to brainstorm how each future scenario impacts the stakeholder

o Start with one future scenario
o Write the stakeholder's name in the centre of the wheel
o Place the headlines around the stakeholder
o Imagine this occurs in the future and brainstorm (and/or research) how each headline would be impacting the stakeholder. These can be positive, negative, or neutral. Brainstorm consequences and combinations until you get a lot of ideas
o When you have a lot, cluster them into themes, and write a summary of each theme to reveal key insights of how the stakeholder will fair in the future scenario

- Repeat the steps above for each future scenario and analyse by looking for any common needs or challenges across the futures that you might nurture or aim to reduce to make the futures better for the stakeholder

# 3.4
# Strategies and lenses for digital

While many of the life-centred strategies relate to physical product design, below are some ideas on how to implement the strategies and lenses for digital products[177]:

## Discover lifecycle

- **Phygital Map**

  o Map the physical components supporting the end-to-end digital lifecycle to understand the peoples, animals, and environments impacted

- **Product anatomy map and Lifecycle map**

  o If your digital product/service has a significant physical component (for example, a SIM card for a mobile phone plan), you could use these maps to make the component more circular

## Align with life

- **Life-centred purpose**

  o Align the business model with global goals

- **Strengthen with transparency**

  o Be transparent to inform and empower users and build their understanding of life-centredness

- **Partner with values**

  o Improve the circularity of your product by partnering with the value web to share and maximise value for all

- **Life-centred culture**

  o 'Life-centre' the business by aligning governance, ownership, finance, and culture with the life-centred purpose

## Design for lifecycle

- **Sustainable and ethical digital**

  o Design with *Sustainable and ethical digital* strategies and advocate for the transition of digital channels to green hosting
  o Embed feedback mechanisms and utilise social platforms to learn continuously, break down silos, connect communities, and drive user-designer-value web collaboration

- **Foster user stewardship**

  o Guide users into more sustainable and respectful behaviour

## Lenses

- **Nature Lens**

  o Draw from nature's information handling and patterns for designing IT systems and AI
  o Use the Golden Ratio tool for visual design

- **People Lens**

  o Use the Power Pixel to know your personal privileges and use them to address injustice, in product, process, supply chain, or workplace
  o Design for pluri-clusivity as the default experience, using the complexity of diversity as inspiration for innovation. For example, allow for user customisation and localisation of experience

- **Regeneration Lens**

  o Take a systems view of your product, business, and 3rd parties to identify resources and communities to renourish

- **Energy Lens**
  - o Power the business and supply chain with efficient renewables and encourage users to do the same

- **Time Lens**
  - o Map future scenarios and test your design ideas to protect future stakeholders

# 3.5
# Methods to get started

Life-centred design's expanded responsibilities, considerations, and interconnectedness might make deciding where to start overwhelming. And with much of the strategies focused on the design of physical products, digital designers may be extra uncertain.

Start by thinking about what your role has direct influence on, for example:

- UX designers can draw on behavioural design to encourage sustainable user behaviour
- Product engineers can map safe and ethical materials to share as guidance for suppliers and manufacturers
- Decision makers can influence which partners the supply chain engages

But anyone can experiment with any strategy and share it back to their team for inspiration. While some strategies and lenses may sound daunting and beyond your responsibility, the designer role is changing to be able to adapt to all levels of problem solving. Experimenting beyond your influence can be an opportunity to build an understanding of lifecycle and systems thinking.

And if no one else in the business is thinking about these connections, someone needs to inspire them—who knows where that may lead you and the business?

Following are suggested methods to get started.

## Experiment

Make starting super easy by picking a random strategy and running through it with your product with the intention of just getting familiar with life-centred thinking.

## Align your purpose

Start with the *Life-centred Purpose* tool to explore how your organisation's and your own purpose might align with global goals.

## One small change

If you're looking to improve circularity of an existing business or product, you might be overwhelmed with a lot of things to do, so try to start with one small change.

- Assess your product/business with the *Lifecycle Map*
- Review the *Strategies* to identify points that need intervention, and consider which ones you have direct influence over to find potential points of leverage
- Apply *Lenses*
- Prioritise the potentials using the 2x2 grid on the *Gather and Prioritise* tool
- Ideate solutions
- Apply a forecasting strategy to test the design's long-term impact and resilience
- Iterate, test, launch, and monitor

## Start small for bigger

Making a long-term, wider-reaching strategy of change can begin with 'incremental innovation' through small changes that compound over time[178]:

- Identify a big problem
- Strategize a solution
- Break it down into small achievable steps
- Solve one at a time using the *One small change* method above
- Monitor the impact and iterate the strategy as needed

## Dual streams

Run dual streams of product iterations and long-term business change.

Test and iterate in small steps to minimise process waste (time and resources), to constantly learn and grow, with a focus on increasing value across the lifecycle rather than just short-term financial profit.

**Case study - Impossible | Bond Touch**

Impossible's approach to make their relationship-focused *Bond Touch* wearable more 'planet centric' was to select SDGs related to the business and translate them into actionable goals.

For example, they translated the following SDGS into these goals:

- *SDG 3 Good health and wellbeing* into 'Support education in building and marinating healthy relationships'
- *SDG 8 Decent work and economic growth* into the 'Ensure good working conditions across the value chain'
- *SDG 13 Climate action* into 'Minimise carbon emissions'
- *SDG 5 Gender equality* into 'Foster gender equality in relationships'
- *SDG 10 Reduced inequalities* into 'Empower and promote all kinds of relationships'
- *SDG 12 Responsible consumption and production* into 'Work towards circularity and foster responsible consumption'

**From there, they created two streams of work.**

In one stream, they defined the new planet centric purpose of the business, expanding the business goal of 'connecting people' to include a focus on healthy relationship building through providing support and education about maintaining balanced relationships.

Using this renewed purpose as a guide, they ideated new product features.

In another stream, they analysed the sustainability of the business from the three perspectives of environmental, social, and economic in terms of what they were already doing and how they could improve. This led to ideating how to make their product more inclusive, and their product lifecycle and technology to transition to more circular operations.

From these ideas, they created a transition strategy and roadmap.

**Let's break that down into steps**

Assess the lifecycle to identify:

- Short term value/product enhancement

  - Value assessment
  - Expand it out to all levels (individual, groups, environment)
  - Use this to define a life-centred vision
  - Ideate ways to spread that vision via the product/service
  - Create a roadmap and metrics

- Short term sustainability fixes to the broken loops under your control
- Long term business remodel

  - Learn about the entire supply chain
  - Analyse for social, economic, and environmental impacts
  - Identify actions to be done for inside and outside the business
  - Use these insights to develop a more sustainable and just business model
  - Create a roadmap for the transitions and set metrics

**You can view Impossible's amazing life-centred work here:**

*miro.com/app/board/o9J_lWvtHAM=*

## Explore partners

Use the *Lifecycle Map* to identify all value web partners:

- Map them on an actual map to see them from a location perspective
- Learn more about them and see them through the *People Lens*
- Draw a non-linear systems view of your lifecycle to look at it from a different angel to identify any unseen or under-utilised relationships or leverage points

## Look outside

Look outside your product or business and find another design problem in the world to solve:

- As an individual looking to invest your time, use the *Lifecycle Map* tool to assess a physical or digital product that:

  - You love or loathe
  - Has excess packaging or waste
  - Is made locally
  - Unfairly advantages or marginalises people
  - Fosters destructive or addictive habits
  - Any product that might need some circular, pluri-clusive, and responsible thinking

- Apply a *Lens*
- Define a *Strategy* and ideate solutions
- Use the *Product To Service* tool to explore the product as a service

# What's next?

How we design and how we teach design are in flux.

By the time this book is published, life-centred design may be more commonly referred to as 21st Century design, planet centric design, or Doughnut for Product Design.

Perhaps, to foster a true pluriverse, there should be many life-centred design frameworks.

Perhaps, in a transformative future scenario, extracting new virgin materials becomes banned and product design becomes completely decentralised and evolves into a citizen-led remaking and recycling of existing materials.

Regardless of what we call life-centred design or how we define it, these unpredictable times are demanding the expansion of human-centred design and change in the way many of us live.

Whether we mitigate the effects of climate change enough or are forced to adapt more than we're expecting, science and lived experience show serious changes with significant impacts are already happening. Supply chains and resources are being constantly challenged, making a sustainable, regenerative, pluri-clusive, and responsible approach to life-centred thinking a must, not a novelty.

Education will need to adapt also as changes happen faster and more often. Educators of climate design already lament how little is taught in design schools, and they worry about education not adapting fast enough[179].

Fortunately, life-centred design curriculums are already emerging:

- Life-centred thinking is being developed from an educational perspective by *The Future of Design Education Initiative*[180]
- *interaction-design.org* offer the life-centred course *21st Century Design* with Don Norman181
- Vision-driven design researcher Masaki Iwabuchi has been developing a design training curriculum that includes the above practices and more[182]

With the current and future state of the world growing more unpredictable, individual designers, business, supply chains, and consumers can no longer ignore their part in the systemic flow of cause and effect.

In a distributed future, the boundary between the designer and consumer disappears to create the citizen designer, and with autonomous design, communities practice the design of themselves[183]. Therefore, it could be argued that life-centred thinking should be taught not only in design schools, but in all schools. Circular design, systems mapping, pluriversal design, biomimicry—the awareness of our relationship with each other and the planet that these frameworks spark should be experienced by all of us, especially today's children who have a very different world awaiting them in the near futures.

And this means the role of the designer would change.

The 21st-century designer would become more of a facilitator of creation, by merging their academic skills and knowledge with local, traditional, and pluriversal perspectives.

The core skills a life-centred designer could then be:

- Identify and gather the right people
- Effectively lend themselves, their privileges, and their knowledge to the people they design with
- Facilitate participatory design where the people design their own solutions

Don Norman argues that today's designers already have the foundational skills needed to tackle global issues[184]. Life-centred design expands and re-contextualises those skills to enable designers to implement design in a wider array of cultural, political, and socioeconomic situations, and perhaps forming a reserve army of designers for sustainable and just futures[185].

While life-centred design challenges deeply ingrained ideas of what is value, wealth, success, beauty, fairness, and equality, we must be careful life-centred changes don't cause rebounds.

Therefore, more designers need to be practicing life-centred thinking now, making mistakes and experimenting with its flaws, while adopting a *protopian* mindset to be always improving.

Take what is presented in this guide and mix it up with all the other emerging frameworks, technologies, players, and challenges not mentioned. Create your own framework, share it back for more remixing and sharing, and by doing so decentralise, distribute, and enrich its evolution.

On an individual and personal level, life-centred design can unlock ineffective dreams and hopes paralysed by eco- and future-anxieties by transforming them into active hope[186] by giving you the skills and mindsets to respond to the unpredictable in tangible, practical, and measurable ways.

In 1985, in *Design For The Real World*, Papanek acknowledged the difficulty of championing personal values while maintaining employment, but he called to designers to make the decision to be on the side of social good or not by choosing what they design for[187].

While some designers have the luxury to choose roles so freely, many are still dominated by the need to sustain oneself in a society that provides for these needs mostly via monetary payment. Perhaps when ESG commitments are more ubiquitous to hold businesses accountable, designers will be more empowered to align their craft with their values.

Until then, designers can start their journey by taking the time to:

- Know their values, privileges, assumptions, and biases to use in positive ways
- Plan a transition to align work and values
- Learn, experiment, and share
- Invest their own time in projects of personal interest

## Know values, privileges, assumptions, and biases

Use the *Power Pixel* as a personal tool, and use this insight to define the most suitable action 'mode' that best aligns with your personal strengths and weakness. Consider roles such as:

- **Driver**—Tackling business models head-on to make them more life-centred
- **Scout**—Planting seeds of regeneration from the ground up
- **Agent**—Aligning with a more life-centred business to nurture them into thriving

## Activate the transition

Aligning work with our true values is a journey, and it may take a long time to really get it right. And it might never stop evolving as the world and us keep changing. If you don't feel aligned with the work you are in, here are some useful things to remember:

- Aligning our working life with personal values is a journey that takes time, so it's ok to be out of alignment for a while
- But there is no transition without having a plan and actively working on it:
  - Think of yourself as a business—know your value proposition, what value you bring, your strengths and weakness
  - Think of work as 'personal purpose'—know your personal values, what you really care about, what do you want to see greatly improved in the world, and what things you want to stop
  - Consider what SDGs your value and values most relate to or can contribute to
  - Do some research and reach out to companies that match your values
  - Join hubs of similar mindsets, like *climatedesigners.org*

- Champion life-centredness wherever you are now in whatever way you can to foster momentum

## Learn, experiment, and share

Remember, this guide is just a stepping-stone into the world of life-centred thinking. Use the links at *https://medium.com/@damienlutz/the-life-centred-design-guide-resources-70ced59f992f* to further your learning about the supporting design practices, the strategies and lenses, and the various approaches.

## Stewarding your spare time

In the meantime, while the mass of profit-focused projects is slowly infused with more life-centred thinking, designers can nurture this movement to shift power by investing their own personal time in designing for causes aligned with their values. Steward your spare time to focus life-centred design thinking on causes you're passionate about, that really need help, or that never receive design thinking.

Collaborating and sharing your efforts also generates discussions that enrich social, scientific, technological, and environmental progress.

It doesn't matter if you solve a problem or not, it matters that you shift some of your design thinking superpower from growth-focused to thriving-focused projects, share your efforts with others to continue the discussion.

\*\*\*

If you enjoyed this guide and the resources, please leave a review on the site where you made the purchase. Reviews help self-funded books like *The Life-centred Design Guide* to be found.

Thank you,

Damien.

# Keep this book circular

As a responsible prosumer, you now steward the resources of this book. Here's a few tips on how to keep it circular.

## Maintain and repair

See *wikihow.com/Repair-a-Paperback-Book* for ways to:

- Clean the book
- Insert loose pages
- Repair torn pages
- Reattach a cover
- Fix the binding

Just make sure you use an *environmentally friendly craft glue*, ones with low VOC (volatile organic compounds), are petrochemical free, and/or water-based.

## Don't want to keep it anymore?

Pass the book on to someone who can use it:

- A friend
- A school
- Drop it in at a local library
- Pop it in a street library box

## Can't give it away?

Paperbacks can generally be recycled, so check your local rules to make sure you can drop it in the paper recycling bin.

# Other books by the author

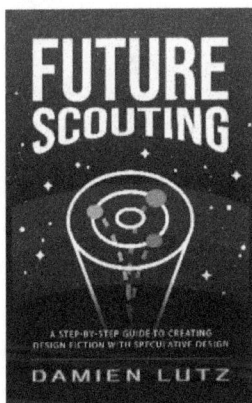

Future Scouting is a fun and practical step-by-step guide to designing fantastic and thought-provoking design fiction prototypes to inspire better tomorrows—an exciting practice known as speculative design. This book includes downloadable worksheets you can use to:

- Catch a signal of emerging change
- Design a future invention
- Ideate a key scenario
- Extrude a Hero and future world
- Compile your artefacts into a shareable prototype

Damien Lutz is a Sydney-based UX/UI Designer, researcher, teacher, and author of *Future Scouting*, a guide to using speculative design to explore the potential impacts of future technology

futurescouting.com.au

damienlutz.com.au

medium.com/@damienlutz

twitter.com/_the_future

# Citations

1 Lin Ph.D, Wambersi MSc, Wackernage, Ph.D.; Estimating the Date of Earth Overshoot Day 2021, 2021; Global Footprint Network, US

2 IPCC, Sixth Assessment Report, 2022, https://www.ipcc.ch/assessment-report/ar6/

3 Might.Design, Future Design Playbook, 2022,
https://import.cdn.thinkific.com/459579/courses/1278497/FuturesDesignPlaybookMight-211114-114016.pdf

4 https://www.commonground.org.au/learn/knowledge-and-sustainability

5 Hao; AI is sending people to jail—and getting it wrong, 2019;
https://www.technologyreview.com/2019/01/21/137783/algorithms-criminal-justice-ai/

6 Pilling; 5 ways GDP gets it totally wrong as a measure of our success, 2018;
https://www.weforum.org/agenda/2018/01/gdp-frog-matchbox-david-pilling-growth-delusion/]

7 Steffen; The nine planetary boundaries, 2015; https://www.stockholmresilience.org/research/planetary-boundaries/the-nine-planetary-boundaries.html

8 Ipsos; Global Trends 2021: Aftershocks and continuity, 2021; https://www.ipsos.com/en/global-trends-2021-aftershocks-and-continuity

9 Union for Ethical BioTrade; UEBT 2020 Biodiversity Barometer, 2020;
https://static1.squarespace.com/static/577e0feae4fcb502316dc547/t/5faba5647c9a080d1659515b/1605084543908/UEBT+Biodiversity+Barometer+2020.pdf

10 Stakeholder Capitalism: A Manifesto for a Cohesive and Sustainable World, 2020;
https://www.weforum.org/press/2020/01/stakeholder-capitalism-a-manifesto-for-a-cohesive-and-sustainable-world/

11 Raworth; Doughnut Economics, 2017; Random House.

12 Raworth; Doughnut Economics, 2017; Random House.

13 United Nations; Sustainable Development Goals, 2015; https://sdgs.un.org/goals

14 Fanning, O'Neill, Hickel, Roux; The social shortfall and ecological overshoot of nations, 2021;
https://www.nature.com/articles/s41893-021-00799-z

15 Circle Economy; The Circularity Gap Report, 2022; https://www.circularity-gap.world/2022#Download-the-report

16 The Guardian; Trauma, dislocation, pollution: why Māori leaders want control of the South Island's water,2021;
https://www.theguardian.com/world/2021/dec/25/trauma-dislocation-pollution-why-maori-leaders-want-control-of-the-south-islands-water

17 Apple Inc; Environmental Progress Report, 2019;
https://www.apple.com/environment/pdf/Apple_Environmental_Progress_Report_2020.pdf

18 Papanek; Design For The Real World, 2019; Thames & Hudson.

19 Design For The 21st Century with Don Norman; https://www.interaction-design.org/courses/design-for-the-21st-century

20 https://circulab.com/who-we-are/

21 Nike Inc.; Sustainable Materials and Innovation; https://www.nike.com/sustainability/materials

22 Kneese, Allen V; The Economics of Natural Resources". Population and Development Review, 1988; doi:10.2307/2808100. JSTOR 2808100.]

23 Ellen MacArthur Foundation; Circular economy introduction; https://ellenmacarthurfoundation.org/topics/circular-economy-introduction/overview

24 Ritzén, Sandström; Barriers to The Circular Economy – Integration of Perspectives and Domains, 2017; https://www.sciencedirect.com/science/article/pii/S221282711730149X

25 Zink, Geyer; Circular Economy Rebound, 2017

26 Sorin, Fabrice; 2021; Circulab

27 Zink, Geyer; Circular Economy Rebound, 2017

28 Institute for Human Centred Design; https://www.humancentereddesign.org/inclusive-design/principles

29 Microsoft Corporation; Inclusive Design, 2018; https://www.microsoft.com/design/inclusive

30 Inclusive Design Principles; https://inclusivedesignprinciples.org/

31 Rohles; More diversity for better experiences, 2020; https://rohles.net/en/articles/diversity-user-experience

32 Buolamwini; Artificial Intelligence Has a Problem With Gender and Racial Bias. Here's How to Solve It, 2019; https://time.com/5520558/artificial-intelligence-racial-gender-bias/

33 Tester; Queering the Future: How LGBTQ Foresight Can Benefit All, 2021; Long Now; https://longnow.org/seminars/02021/jan/26/queering-future-how-lgbtq-foresight-can-benefit-all/

34 Onafuwa, 2021

35 TakingITGlobal Sustainable Development Goal Inclusive Design Toolkit; https://research.tigweb.org/cristian/sdgid/

36 Noel; Envisioning a Pluriversal Design Education, 2020; PIVOT2020; https://www.youtube.com/watch?v=OcXI8erWgcA

37 Leitão, Noel; DRS2020 Editorial: Pluriversal Design SIG, 2020; https://www.researchgate.net/publication/345376444_DRS2020_Editorial_Pluriversal_Design_SIG/citation/download

38 Kothari, Ashish; Salleh, Ariel; Escobar, Arturo; Demaria, Federico; Acosta, Alberto. Pluriverse: A Post–Development Dictionary, 2019; AuthorsUpFront, Tulika Books, New Delhi.

39 Linehan; Modernity, 2009; International Encyclopedia of Human Geography; https://www.sciencedirect.com/topics/earth-and-planetary-sciences/modernity

40 Kothari, Ashish; Salleh, Ariel; Escobar, Arturo; Demaria, Federico; Acosta, Alberto. Pluriverse: A Post–Development Dictionary, 2019; AuthorsUpFront, Tulika Books, New Delhi

41 GSDR 2023 Call for inputs, 2021; https://sdgs.un.org/news/call-inputs-global-sustainable-development-report-2023-34347

42 Kothari, Ashish; Salleh, Ariel; Escobar, Arturo; Demaria, Federico; Acosta, Alberto. Pluriverse: A Post–Development Dictionary, 2019; AuthorsUpFront, Tulika Books, New Delhi.

43 Manitoba Education and Training ; Education for a Sustainable Future: A Resource for Curriculum Developers, Teachers, and Administrators, 2000

44 Whitewashed Hope: A Message from 10+ Indigenous Leaders and Organizations, 2020; https://www.culturalsurvival.org/news/whitewashed-hope-message-10-Indigenous-leaders-and-organizations

45 Onafuwa; What inclusion really means | A pluriversal approach to user experience, 2020;

https://www.ecuad.ca/events/what-inclusion-really-means-a-pluriversal-approach-to-user-experience

46 Rogal; Embracing Many Worlds: the Wixárika calendar, 2020; https://taylor.tulane.edu/pivot/contributions/wixarika-calendar/

47 Larsen; Queer Futures, 2021; https://medium.com/copenhagen-institute-for-futures-studies/queer-futures-with-jason-tester-43c4e65cdfd1

48 Pivot 2021 Conference, 2021; https://pivot2021conference.com/recordings/

49 Taboada, Turner; Narratives and Alternate Futures| Rolling Stones: Dismantling, reassembling and reimagining possible tools through collaborative story-making approaches, 2012; https://vimeo.com/582134090/f6edcbd940

50 Bisht, Decolonizing our Future Through Inclusive Storytelling, 2019; PRIMER19; https://vimeo.com/348308137

51 Escobar; Designs for the Pluriverse (New Ecologies for the Twenty-First Century), 2018; Duke University Press.

52 Graham; Indigenous Perspectives About Quality of Life, 2021; New Economy Network Australia;

https://www.youtube.com/watch?v=rXH2IWGl338

53 Graham; Indigenous Perspectives About Quality of Life, 2021; New Economy Network Australia;

https://www.youtube.com/watch?v=rXH2IWGl338

54 Escobar; Designs for the Pluriverse (New Ecologies for the Twenty-First Century), 2018; Duke University Press.

55 Goodwill; A Social Designer's Field Guide to Power Literacy, 2020; https://www.kl.nl/en/publications/a-social-designers-field-guide-to-power-literacy/; Kennisland

56 Heke, Rees, Swinburn, Waititi, Stewart; Systems Thinking and Indigenous systems: native contributions to obesity prevention, 2018; https://journals.sagepub.com/doi/full/10.1177/1177180118806383

57 Common Ground; First Nations Systems thinking; https://www.commonground.org.au/learn/first-nations-systems-thinking

58 Systems Innovation; Circular Economy Toolkit, 2021; https://www.systemsinnovation.io/post/circular-systems-toolkit

59 Meadows; Thinking in Systems: A Primer; 2008; Chelsea Green Publishing

60 Norman; What is Systems Thinking?; https://www.interaction-design.org/literature/topics/systems-thinking

61 Norman; What is Systems Thinking?; https://www.interaction-design.org/literature/topics/systems-thinking

62 Bouganim; A case for Systemic Design, 2020; https://uxplanet.org/a-case-for-systemic-design-5a9465b870fa; Medium

63 Systems Innovation; Circular Economy Toolkit, 2021; https://www.systemsinnovation.io/post/circular-systems-toolkit

64 Bouganim; A case for Systemic Design, 2020; https://uxplanet.org/a-case-for-systemic-design-5a9465b870fa; Medium

65 La; Why designers should find the balance between systems thinking and design thinking?, 2019;

https://tylerla.medium.com/why-designers-should-find-the-balance-between-systems-thinking-and-design-thinking-efdb57b9949f; Medium

66 Design Council; Beyond Net Zero: A Systemic Design Approach, 2021;

https://www.designcouncil.org.uk/resources/guide/beyond-net-zero-systemic-design-approach

67 Meadows; Thinking in Systems: A Primer, 2008; Chelsea Green Publishing

68 Meadows; Thinking in Systems: A Primer, 2008; Chelsea Green Publishing

69 https://www.youtube.com/watch?v=8rU2FWT5Koc

70 Johnson; Direct mortality of birds from anthropogenic causes, 2015

71 Woebken, Okada; Animal Superpowers, 2007; https://chriswoebken.com/ANIMAL-SUPERPOWERS

72 Chavarria; Interface masks, 2020; https://ciid.dk/education/portfolio/idp20/courses/final-projects/projects/interface/

73 Imagination Lancaster; Interspecies Toolkit; https://www.interspeciesdesign.co.uk/

74 Imagination Lancaster; Interspecies Activity Cards; https://www.interspeciesdesign.co.uk/interspecies-activity-cards/

75 Hirskyj-Douglas, Lucero; On the Internet, Nobody Knows You're a Dog... Unless You're Another Dog, 2019; https://dl.acm.org/doi/fullHtml/10.1145/3290605.3300347

76 Douglas et al, Jukan et al

77 Wirman, Zamansky; Playful Animal-Computer Interactions: a Framework, 2015; http://creativegames.org.uk/modules/Intro_Game_Studies/Wirman_Zamansky_Playful_Animal-Computer_Interactions-2015.pdf

78 Hook; A Report on the First International Workshop on Research Methods in Animal-Computer Interaction, 2017; https://www.cryptoludology.com/?p=550

79 Armstrong , Toribio, Whyman; Distributed Design Book, 2020; https://distributeddesign.eu/resources/

80 Freitag, Berners-Lee, Widdicks, Knowles, Blair, Friday; The real climate and transformative impact of ICT: A critique of estimates, trends, and regulations, 2021; Science Direct

81 https://coderwall.com/p/57imrw/common-fonts-for-windows-mac

82 Capgemini; How sustainability is fundamentally changing consumer preferences, 2020; https://www.capgemini.com/wp-content/uploads/2020/07/20-06_9880_Sustainability-in-CPR_Final_Web-1.pdf

83 Stahel; The Circular Economy: A User's Guide, 2019; Routledge, US

84 Stockholm Resilience Center; Helping design smarter urban green spaces; https://www.stockholmresilience.org/research/research-news/2021-11-18-helping-design-smarter-urban-green-spaces.html

85 https://healabel.com/

86 https://medium.com/@good.gang/challenging-uber-eats-the-app-that-could-promote-planet-friendly-eating-habits-7e2c94f07d

87 https://www.greenchoicenow.com/

88 Strömberg, Selvefors, Renström; Mapping out the design opportunities: pathways of sustainable behaviour, 2015; https://www.tandfonline.com/doi/full/; International Journal of Sustainable Engineering

89 Brignall; Deceptive Design; https://www.darkpatterns.org/types-of-dark-pattern

90 Neuenschwander, Berdichevsky; Toward an Ethics of Persuasive Technology, 1999; Association for Computing Machinery, US

91 Combs PhD, Brian; Digital Behavioural Design, 2018; https://s3.amazonaws.com/arena-attachments/2150295/ecc52e80b8852ed927eba5a66ec3b44e.pdf

92 Lidman, Renström; How to design for sustainable behaviour?, 2011; Chalmers University of Technology

93 Stahel; The Circular Economy: A User's Guide, 2019; Routledge, US

94 Stockholm Resilience Centre; Stewardship and transformative futures; https://www.stockholmresilience.org/research/research-themes/stewardship-transformation.html

95 Stockholm Resilience Centre; Stewardship and transformative futures; https://www.stockholmresilience.org/research/research-themes/stewardship-transformation.html

96 Ottmann, PhD; Sustainability & Reconciliation: Indigenous Perspective, 2019; https://www.saskwastereduction.ca/assets/upload/pdf/events/reforum2019/ottmann-presentation-april-2019-5cb4e46a421ef.pdf

97 McClure; Trauma, dislocation, pollution: why Māori leaders want control of the South Island's water, 2021; https://www.theguardian.com/world/2021/dec/25/trauma-dislocation-pollution-why-maori-leaders-want-control-of-the-south-islands-water; The Guardian

98 Kimmerer; Braiding Sweetgrass: Indigenous Wisdom, Scientific Knowledge and the Teachings of Plants, 2020; Penguin Random House, UK

99 Quintus et al. 2019

100 Ticktin et al. 2018

101 Winter, Lincoln, Berkes, Alegado, Kurashima, Frank, Pascua, Rii, Reppun, Knapp, McClatchey, Ticktin, Smith, Franklin, Oleson, Price,McManus, Donahue, Rodgers, Bowen, Nelson, Thomas, Leon, Madin, Rivera, Falinski, Bremer, Deenik, Got III, Neilson, Kano, Olegario, Nyberg, Kawelo, Kotubetey, Kukea-Shultz, Tonnon; Ecomimicry in Indigenous resource management: optimizing ecosystem services to achieve resource abundance, with examples from Hawai'i, 2020; https://www.ecologyandsociety.org/vol25/iss2/art26/#formsofecomi8; Ecology and Society

102 Winter et al. 2018a

103 Jacobi et al. 2017

104 Loke and Leung 2013

105 Baumeister, Ph.D.; Biomimicry Resource Handbook, 2014; Biomimicry 3.8, US

106 Baumeister, Ph.D.; Biomimicry Resource Handbook, 2014; Biomimicry 3.8, US

107 Goss, J. Biomimicry: Looking to nature for design solutions, 2009; https://www.proquest.com/openview/eb78630603853e491f17ba27d8317f72/1; Corcoran College of Art + Design ProQuest Dissertations Publishing,

108 Baumeister, Ph.D.; Biomimicry Resource Handbook, 2014; Biomimicry 3.8, US

109 Biomimicry Institute; https://biomimicry.medium.com/meet-the-newest-young-innovators-in-biomimicry-a466cd3d9649, 2022; Medium

110 Benyus; Biomimicry: Innovation Inspired by Nature, 2003; HarperCollins, US

111 Marshall; The Theory and Practice of Ecomimicry, 2007; https://www.researchgate.net/publication/306399880_The_Theory_and_Practice_of_Ecomimicry

112 Orr; Ecological Literacy: Education and the Transition to a Postmodern World, 1991; SUNY Press, US

113 Dwyer; The Ultimate Guide To The Genius Of Place, 2018; https://synapse.bio/blog/ultimate-guide-to-genius-of-place

114 Baumeister, Ph.D.; Biomimicry Resource Handbook, 2014; Biomimicry 3.8, US

115 Bio-inspired, Another Approach, 2020; https://www.cite-sciences.fr/fr/au-programme/expos-permanentes/les-expositions/bio-inspiree/lexposition/la-voie-de-la-bio-inspiration/

116 Larsen; What Is 'Futures Literacy' and Why Is It Important?, 2020; https://medium.com/copenhagen-institute-for-futures-studies/what-is-futures-literacy-and-why-is-it-important-a27f24b983d8; Medium

117 Norman; Principles of Human-Centered Design (Don Norman), 1998-2022; https://www.nngroup.com/videos/principles-human-centered-design-don-norman

118 Kelley, Kelley; Creative Confidence, 2013; William Collins, UK

119 Positive; The Purpose Toolkit, 2022; https://www.makeapositiveimpact.co/changemaker-resources

120 Positive; The Positive Compass, 2022; https://www.makeapositiveimpact.co/positive-compass

121 SPACE10; Regenerative by Design, 2021; https://space10.com/regenerative-by-design

122 Manitoba Education and Training ; Education for a Sustainable Future: A Resource for Curriculum Developers, Teachers, and Administrators, 2000; https://www.edu.gov.mb.ca/k12/cur/socstud/frame_found_sr2/tns/tn-41.pdf

123 Vasconcellos, de Fraguier; The Positive Handbook for Regenerative Business - Stephen Vasconcellos & Niels de Fraguier.pdf, 2021; https://www.makeapositiveimpact.co/changemaker-resources-handbook

124 Wahl, Can regenerative economics & mainstream business mix?, 2019; https://medium.com/activate-the-future/can-regenerative-economics-mainstream-business-mix-ef2f8aafa8d4

125 Positive; ETHICAL SOURCING; Ending Modern Slavery, 2022; https://www.makeapositiveimpact.co/

126 Schultz, 2019; Cited in https://acuads.com.au/wp-content/uploads/2020/05/Kelly-Meghan_2019.pdf

127 Raworth; Doughnut Economics, 2017; Random House.

128 Patagonia; Social Responsibility; https://www.patagonia.com/social-responsibility/

129 Goodwill; A Social Designer's Field Guide to Power Literacy, 2020; https://www.kl.nl/en/publications/a-social-
designers-field-guide-to-power-literacy/; Kennisland

130 https://doughnuteconomics.org/about-doughnut-economics

131 SYSTEMIQ; Everything-as-a-service, 2021; https://www.systemiq.earth/wp-content/uploads/2021/11/XaaS-
ExecutiveSummary.pdf

132 Rajagopalan, Dunnett, Deeken, Born, Brass, Stemp, Morgenstern, Naeve, Erlwein; Understanding the SDGs in
sustainable investing, 2018; https://www.berenberg.de/fileadmin/web/asset_management/news/esg-
news/SDG_understanding_SDGs_in_sustainable_investing.pdf

133 Vasconcellos, de Fraguier; The Positive Handbook for Regenerative Business, 2021;
https://www.makeapositiveimpact.co/changemaker-resources-handbook

134 Philips; Our Purpose, 2022:https://www.philips.com/a-w/about/environmental-social-governance/our-purpose

135 Positive; The Purpose Toolkit, 2022; https://www.makeapositiveimpact.co/changemaker-resources

136 Positive; The Positive Compass, 2022; https://www.makeapositiveimpact.co/positive-compass

137 Union For Ethical Biotrade; UEBT Biodiversity Barometer 2020;
https://static1.squarespace.com/static/577e0feae4fcb502316dc547/t/5faba5647c9a080d1659515b/1605084543908/UEBT+
Biodiversity+Barometer+2020.pdf

138 Haynam; The Transparency Movement: What It Is, Why It's Important And How to Get Involved, 2015;
https://buffer.com/resources/transparency-movement

139 https://www.globalreporting.org/how-to-use-the-gri-standards

140 Stahel; The Circular Economy: A User's Guide, 2019; Routledge, US

141 Vasconcellos, de Fraguier; The Positive Handbook for Regenerative Business, 2021;
https://www.makeapositiveimpact.co/changemaker-resources-handbook

142 Vasconcellos, de Fraguier; The Positive Handbook for Regenerative Business, 2021;
https://www.makeapositiveimpact.co/changemaker-resources-handbook

143 Symbiotic Networks of Bio-Waste Sustainable Management; A good example of Industrial Symbiosis, 2020;
https://symbiosisproject.eu/a-good-example-of-industrial-symbiosis

144 SCALER Project; Introduction to industrial symbiosis, 2020; https://www.youtube.com/watch?v=7daVvUsvBuc

145 Vasconcellos, de Fraguier; The Positive Handbook for Regenerative Business, 2021;
https://www.makeapositiveimpact.co/changemaker-resources-handbook

146 https://doughnuteconomics.org/about-doughnut-economics

147 Australian Human Rights Commission; Cultural diversity in the workplace, 2015;
https://humanrights.gov.au/about/news/speeches/cultural-diversity-workplace-0

148 Ordorica; The How And Why Of Building A Diverse Workforce, 2021;
https://www.forbes.com/sites/forbesbusinesscouncil/2021/07/26/the-how-and-why-of-building-a-diverse-workforce/

149 Academy of Management; The Effects of Racial Diversity Congruence between Upper Management and Lower
Management on Firm Productivity, 2021; https://journals.aom.org/doi/10.5465/amj.2019.0468

150 Professor Messer; Device Disassembly Best Practices - CompTIA A+ 220-901 - 4.5, 2017;
https://www.youtube.com/watch?v=g8I7lPO_wFY

151 De Fazio, Bakker, Flipsen, Balkenende; The Disassembly Map: A new method to enhance design for product
repairability, 2021; https://www.sciencedirect.com/science/article/pii/S0959652621027608; ScienceDirect

152 Maartens, De Fazio; Making circular innovation work
Design for disassembly, 2021; https://www.engineeringsolutions.philips.com/app/uploads/Design-for-disassembly-20211014.pdf: Koninklijke Philips N.V.

153 Positive; The Purpose Toolkit, 2022; https://www.makeapositiveimpact.co/changemaker-resources

154 Positive; The Positive Compass, 2022; https://www.makeapositiveimpact.co/positive-compass

155 Butoliya; Critical Tools: Design pedagogy for alternative ways of making futures, 2021; https://pivot2021conference.com/recordings; Pivot 2021

156Butoliya; Critical Tools: Design pedagogy for alternative ways of making futures, 2021; https://pivot2021conference.com/recordings; Pivot 2021

157 Acaroglu; Quick Guide to Circular Economy Business Strategies, 2020; https://medium.com/disruptive-design/quick-guide-to-circular-economy-business-strategies-b3d6a000facf; Medium

158 Papanek; Design For the Real World, 2020; Thames & Hudson, UK

159 Positive; Ethical Sourcing; Ending Modern Slavery, 2022; https://www.makeapositiveimpact.co/changemaker-resources

160 Cradle to Cradle Products Innovation Institute Inc.; *What is Cradle to Cradle Certified*, 2020; https://www.c2ccertified.org/get-certified/product-certification

161 Ellen MacArthur Foundation; Safe & Circular: Material Selection, Regrettable Substitution, 2018; https://www.youtube.com/watch?v=KugNkZeBrRg&t=2s

162 Noe; A Concrete Block Alternative Made from Difficult-to-Recycle Plastic; https://www.core77.com/posts/111436/A-Concrete-Block-Alternative-Made-from-Difficult-to-Recycle-Plastic; Core77

163 https://plasticwhale.com/

164 Maartens, De Fazio; Making circular innovation work—Design for disassembly, 2021; https://www.engineeringsolutions.philips.com/app/uploads/Design-for-disassembly-20211014.pdf: Koninklijke Philips N.V.

165 Professor Messer; Device Disassembly Best Practices - CompTIA A+ 220-901 - 4.5, 2017; https://www.youtube.com/watch?v=g8I7lPO_wFY

166 De Fazio, Bakker, Flipsen, Balkenende; The Disassembly Map: A new method to enhance design for product repairability, 2021; https://www.sciencedirect.com/science/article/pii/S0959652621027608; ScienceDirect

167 Maartens, De Fazio; Making circular innovation work
Design for disassembly, 2021; https://www.engineeringsolutions.philips.com/app/uploads/Design-for-disassembly-20211014.pdf: Koninklijke Philips N.V.

168 Marks et al., 1993

169 Circle Lab; Incorporate digital technology; https://knowledge-hub.circle-lab.com/frameworks/9/162?n=Incorporate-digital-technology; Circle Economy

170 EcoDesign Circle; Circular Business Models, 2018; https://sustainabilityguide.eu/methods/circular-business-models

171 Strömberg, Selvefors, Renström; Mapping out the design opportunities: pathways of sustainable behaviour, 2015; https://www.tandfonline.com/doi/full/; International Journal of Sustainable Engineering

172 Snzel; 2020; https://medium.com/user-experience-design-1/your-next-persona-will-be-animal-tools-for-environment-centered-designers-c7ff96dc2b17

173 Goodwill; A Social Designer's Field Guide to Power Literacy, 2020; https://www.kl.nl/en/publications/a-social-designers-field-guide-to-power-literacy/; Kennisland

174 Goodwill; A Social Designer's Field Guide to Power Literacy, 2020; https://www.kl.nl/en/publications/a-social-designers-field-guide-to-power-literacy/, Kennisland

175 Matchbox Studio; Starting From Place, 2020; https://www.matchboxstudio.com.au/starting-from-place

176 Imagination Lancaster; Interspecies Design; https://www.interspeciesdesign.co.uk/interspecies-activity-cards

177 Määttä; Circular goes digital;

https://www2.deloitte.com/content/dam/Deloitte/fi/Documents/risk/Circular%20goes%20digital.pdf. Deloitte

178 Norman; 21st Century Design; https://www.youtube.com/watch?v=7FJNsqoC4tI; NNgroup

179 Zimmermann; Sustainable Design Should Be Foundational;

https://www.climatedesigners.org/edu/climifypodcast/sustainable-design-should-be-foundational; Climate Designers Org

180 Future of Design Education; https://www.futureofdesigneducation.org/

181 21st Century Design with Don Norman; https://www.interaction-design.org/master-classes/21st-century-design-with-don-norman

182 Iwabuchi; Emerging Design Attitudes: Speculative, Transitional, and Pluriversal Design, 2020;

https://uxplanet.org/design-attitudes-for-this-century-speculative-transitional-and-pluriversal-design-fb55c9d401e6;

Medium

183 Escobar; Designs for the Pluriverse (New Ecologies for the Twenty-First Century), 2018; Duke University Press.

184 21st Century Design with Don Norman; https://www.interaction-design.org/master-classes/21st-century-design-with-don-norman

185 Norman; 21st Century Design; https://www.youtube.com/watch?v=7FJNsqoC4tI; NNgroup

186 Active hope; https://www.activehope.info/

187 Papanek; Design For the Real World, 2020; Thames & Hudson, UK